Green Electrical Energy Storage

Green Electrical Energy Storage

Science and Finance
for Total Fossil Fuel Substitution

Gabriele Zini

New York Chicago San Francisco Athens
London Madrid Mexico City Milan
New Delhi Singapore Sydney Toronto

ISBN 978-1-25-964283-8
MHID 1-2-5964283-6

The pages within this book were printed on acid-free paper.

Sponsoring Editor Michael McCabe	**Copy Editor** Cenveo Publisher Services	**Production Supervisor** Lynn M. Messina
Editorial Supervisor Donna M. Martone	**Proofreader** Cenveo Publisher Services	**Composition** Cenveo Publisher Services
Acquisitions Coordinator Lauren Rogers	**Indexer** Gabriele Zini	**Art Director, Cover** Jeff Weeks
Project Manager Srishti Malasi, Cenveo® Publisher Services		

This book is dedicated to our children, to show them that we worked hard to leave them a better world than the one we were given.

Contents

List of Figures

List of Tables

Preface

Some of the most available, cheapest, and powerful renewable energy sources for green electric energy conversion are highly uneven in nature.

Solar photovoltaic plants convert renewable energy during day-time, but fail to do so when solar radiation is low or absent during bad weather conditions or night-time. Wind energy manifests even higher variability than solar photovoltaics, as well as posing greater difficulties in correct forecasting of wind turbine output.

In both cases, the energy converted from such renewable sources is immediately injected in the grid with the risk of making the electric distribution grid unstable. A national energy system with a large penetration of intermittent renewable energy sources can therefore be extremely difficult, if not impossible, to manage, with the uneconomic curtailment of solar photovoltaic and wind energy sources being the only alternative.

To efficiently deal with the injection of large shares of uneven and unpredictable energy, the distribution networks are starting to upgrade to new demand-side or supply-side management methodologies, taking advantage of new smart-grid devices. Anyway, real-time management is likely not sufficient to cope with the new challenges tied to the substitution of fossil fuels, since there must be in place the possibility to postpone the use of energy should unbalances between energy supply and demand occur. Robust and stable power grids must therefore be capable of distribute energy both in space as well as in time.

Energy storage is then the key to enable decisions about when to use the energy converted from natural sources that cannot be time-controlled. Energy storage systems for large-scale applications are therefore considered an important medium-term development of energy networks.

Storage can be employed for power-intensive applications, where power is delivered for relatively short periods of time, or energy-intensive applications, where energy is delivered to the load for relatively long periods of time. The number of available applications made possible through the adoption of energy storage technologies are a myriad and all of them, taken singly or combined, provide enormous benefits to the energy infrastructure.

The purpose of this book is to provide the knowledge of the science and technology of electric energy storage together with *real-life* financial and contractual frameworks to effectively implement energy storage projects, moving from the idea to *real-life* operations.

The introduction to the book discusses the topic, highlighting the reasons why energy storage system are gaining relevance in the energy markets, and provides the generic concepts and basic information on energy storage technologies. The book is further divided into two parts; chapters 1 to 4 describe the science and technology of different types of energy storage; chapter 5 closes the first part by discussing design and interfacing of storage systems. Part 1 is the "Making It" of energy storage projects; without this knowledge, no energy storage projects can be designed. In part 2, chapter 6 provides the reader with the tools to analyze the economic and financial aspects leading to decision-making for the implementation of large-scale energy projects, as well as the risks that can hamper their successful development and management; chapter 7 outlines the legal and contractual framework necessary to negotiate and contract the construction and operations of energy plants; chapter 8 shows how to analyze, negotiate, and obtain financing for energy projects. This time, part 2 is the "Making It Happen"; without this knowledge, energy projects could result very difficult to be successfully developed and built. At the end of every chapter the relevant bibliography is listed for the reader who wishes to further explore the topics covered in the chapters.

The technologies that are discussed in this book for green electricity storage employs electrochemical, chemical, electrical, and flywheel mechanical devices. This book is not about mechanical or thermal energy technologies that can be difficult to be applied to different scales of power or to non-stationary uses. Electric energy can be stored in hydroelectric plants, in compressed air plants, in gravel plants as mechanical energy. Their use is anyway almost restricted to very large power plants, and can become difficult to be scaled down to smaller power sizes (under the 1 MW threshold) with non-stationary, or stationary applications for residential, commercial, or industrial power becoming very limited. Thermal energy storage is not discussed in this book although, for instance, solar thermodynamic plants can store thermal energy to be converted back into electricity at a later time. Such plants are as well normally very large and difficult to scale down due to excessive losses in efficiency when under the same 1 MW limit. Finally, the choice not to include too many specific data about different technologies has been deliberate; it has been indeed preferred not to provide the reader with data that would have been too much technology-dependent and that will be superseded by technological advances or manufacturing changes.

This book has been conceived with the goal to share with the reader not only the science and technology concepts, but also the necessary tools to improve the chances of the implementation of real-life renewable energy storage projects. The underlying certainty is that these plants will improve the quality of life of our next generations, and this belief has made the effort of writing this book worthwhile.

Gabriele Zini

Acronyms

AC	Alternate Current, Activated Carbon
ACE	Area Control Error
ADSCR	Annual Debt Service Cover Ratio
AE	Alkaline Electrolyzer
AFC	Alkaline Fuel Cell
AGC	Automated Generation Control
AHI	Aqueous Ion
ALOP	Advance Loss of Profit (Insurance)
ASCII	American Standard Code for Information Interchange
AVDSCR	Average Debt Service Cover Ratio
BESS	Battery Energy Storage System
BET	Brunauer-Emmett-Teller
B/S	Balance Sheet
BMS	Battery Management System
BOS	Balance of System
BWR	Back-Work Ratio
CAC	Construction Acceptance Certificate
CAES	Compressed Air Energy Storage
CAPEX	Capital Expenses
CAPM	Capital Asset Pricing Model
CCGT	Combined Cycle Gas Turbine
CFADS	Cash Flow Available for Debt Service
CHP	Combined Heat and Power
CIF	Cost, Insurance and Freight
COD	Commercial Operation Date
COGS	Cost of Goods Sold
COP	Coefficient of Performance
CPI	Consumer Price Index
CSC	Current Source Converter
CSP	Concentrated Solar Power
DA	Distribution Agreement
DAP	Delivered At Place
D&A	Depreciation and Amortization
DC	Direct Current
D/E	Debt to Equity

DD	Due Diligence
DL	Double Layer
DOD] Depth of Discharge DOE	Department Of Energy
DPBP	Discounted Pay Back Period
DSCR	Debt Service Coverage Ratio
DSRA	Debt Service Reserve Account
EA	Evolutionary Algorithm
EAR	Erection All Risk (Insurance)
EBIT	Earning Before Interest and Taxes
EBITDA	Earning Before Interest, Taxes, Depreciation and Amortization
ECA	Export Credit Agency
EDL	Electrical Double Layer
EDLC	Electrochemical Double Layer Capacitor
EIA	Environmental Impact Analysis
EET	Energy Efficiency Title
EL	Electrolyzer
EMS	Energy Management System
EOC	End Of Charge
ESR	Equivalent Series Resistance
ESS	Energy Storage System
ESU	Energy Storage Unit
EUR	Euro (currency)
EVA	Economic Value Added
EV	Electric Vehicle
EXW	Ex-Works
FAC	Final Acceptance Certificate
FAS	Free Alongside Ship
FB	Flow Battery
FC	Fuel Cell
FCA	Free-Carrier
FCF	Free Cash Flow
FCFE	Free Cash Flow to the Equity
FCFF	Free Cash Flow to the Firm
FESS	Flywheel Energy Storage System
FERC	US Federal Energy Regulatory Commission
FIT	Feed-In Tariff
FF	Filling Factor
FOB	Free On Board
FRT	Fault Ride Through
FTP	File Transfer Protocol
GA	Genetic Algorithms
GAAP	Generally Accepted Accounting Principles
G&A	General and Administrative
GHG	Greenhouse Gas
GOOSE	Generic Object Oriented Substation Events
GSE	Generic Substation Events
GT	Game Theory

GTP	Gas-to-Power
GUI	Graphical User Interface
HA	Hydrogen Attack
HC	Hydrocarbon
HCV	Higher Calorific Value
HE	Hydrogen Embrittlement
HEV	Hybrid Electric Vehicle
H&S	Health and Safety
HFL	Higher Flammability Limit
HHV	Higher Heating Value
HMI	Human Machine Interface
HTE	High Temperature Electrolysis
HTF	Heat Transfer Fluid
HTS	High Temperature Shift, High Temperature Superconductor
HV	High Voltage
IAS	International Accounting Standards
ICB	Iron-Chromium Battery
ICC	International Chamber of Commerce
ICD	IED Capability Description
IEA	International Energy Agency
IEC	International Electrotechnical Commission
IED	Intelligent Electronic Device
IEEE	Institute of Electrical and Electronics Engineers
IGBT	Insulated Gate Bipolar Transistor
IP	Intellectual Property, Internet Protocol
IPO	Initial Purchase Offer
IPP	Independent Power Producer
IPR	Intellectual Property Rights
IPS	Interface Protection System
IQAR	Identification, Quantification, Analysis, Response
IRR	Internal Rate of Return
IUPAC	International Union of Pure and Applied Chemistry
JVA	Joint Venture Agreement
KBM	Kinetic Battery Model
KERS	Kinetic Energy Recovery System
LA	Lead Acid
LC	Inductance-Capacitance
LCOE	Levelized Cost of Energy
LCV	Lower Calorific Value
LD	Liquidated Damages
LF	Langmuir-Freundlich (equation)
LFL	Lower Flammability Limit
LHV	Lower Heating Value
LIB	Lithium-Ion Battery
LLCR	Loan Life Cover Ratio
LMTD	Log Mean Temperature Difference
LTS	Low Temperature Shift, Low Temperature Superconductor
LV	Low Voltage

LVRT	Low Voltage Ride Through
MAC	Material Adverse Change
MAE	Material Adverse Event
M&A	Mergers and Acquisitions
MCA	Monte Carlo Analysis
MCFC	Molten Carbonate Fuel Cell
MCP	Measure, Correlate, Predict
MIRR	Modified Internal Rate of Return
MMI	Man-Machine Interface
MMS	Manufacturing Message Specification
MOSFET	Metal Oxide Semiconductor Field Effect Transistor
MPPT	Maximum Power Point Tracking
MSB	Most Significant Byte
MWCNT	Multi-Wall Carbon Nano-tube
MV	Medium Voltage
NDA	Non Disclosure Agreement
NBO	Non Binding Offer
NBP	Normal Boiling Point
NFP	Net Financial Position
NPV	Net Present Value
NTP	Notice to Proceed
OCF	Operating Cash Flows
OCV	Open Circuit Voltage
OVRT	Over-Voltage Ride Through
OPEX	Operating Expenses
ORC	Organic Rankine Cycle
OTC	Over The Counter
OTEC	Ocean Thermal Energy Conversion
PAC	Provisional Acceptance Certificate
PAFC	Phosphoric Acid Fuel Cell
PB	Power Block
PBP	Pay-Back Period
PC	Programmable Computer
PCA	Project Collaboration Agreement
PCS	Power Conversion/Control/Conditioning System
PDA	Project Development Agreement
PDE	Partial Differential Equation
PDF	Probability Density Function
PE	Private Equity
PEM	Proton Exchange Membrane, Polymer Electrolyte Membrane
PEMFC	Proton Exchange Membrane Fuel Cell, Polymeric Electrolyte Membrane Fuel Cell
PF	Project Financing
PFR	Plug Flow Reactor
PHES	Pumped Hydro Energy Storage
P&L	Profit and Loss
PLC	Programmable Logic Controller
PLCR	Project Life Cover Ratio

PM	Particulate Matter
PME	Polymeric Membrane Electrolyzer
PMF	Probability Mass Function
PMSM	Permanent Magnet Synchronous Machine
POD	Point of Delivery
PPA	Power Purchase Agreement
PPC	Power Plant Controller
PSO	Particle Swarm Optimization
PTG	Power-to-Gas
PU	Power Unit
PV	Photovoltaic
PWM	Pulse Width Modulation
QOS	Quality of Service
QPA	Quota Purchase Agreement
RAR	Risk Analysis and Response
RC	Resistance-Capacitance
R&D	Research and Development
RE	Renewable Energy
RECS	Renewable Electricity Certificate System
redox	Reduction-Oxidation
REN	Renewable Energy
RES	Renewable Energy Sources
RFB	Redox Flow Battery
RMS	Root Mean Square
ROA	Return on Assets
ROFR	Right of First Refusal
RT	Round Trip (efficiency)
RTU	Remote Terminal Unit
R&W	Representations and Warranties
SCL	Substation Configuration Language
S&P	Standard and Poor's
SHC	Specific Heat Capacity
SHE	Standard Hydrogen Electrode
SHES	Solar Hydrogen Energy System
SA	Simulated Annealing, Simplex Algorithm
SCADA	Supervisory Control And Data Acquisition
SCC	Social Cost of Carbon
SCD	Substation Configuration Description
SPA	Sale and Purchase Agreement
SME	Small Medium Enterprise
SMES	Superconducting Magnetic Energy Storage
SMR	SteaM Reforming
SNG	Synthetic Natural Gas, syngas
SOC	State Of Charge
SOFC	Solid Oxide Fuel Cell
SOH	State of Health
SPA	Share Purchase Agreement
SPE	Solid Polymer Electrolyzer

SPV	Special Purpose Vehicle
SRC	Specific Rated Capacity
SSD	System Specification Description
SSR	Sub-Synchronous Resonance
STATCOM	Static Synchronous Compensator
STOR	Short Term Operating Reserve
STP	Standard Temperature and Pressure
SWCNT	Single Wall Carbon Nano Tube
TCP	Transmission Control Protocol
T&D	Transmission and Distribution
TM	Trademark
TOU	Time of Use
UC	Ultracapacitor
UFLS	Under-Frequency Load Shedding
UPS	Uninterruptible Power Supply
US	United States
USA	United States of America
USD	United States Dollar (currency)
VC	venture Capital
VMD	Virtual Manufacturing Device
VPN	Virtual Private Network
VRB	Vanadium Redox Battery
VRLA	Valve Regulated Lead-Acid
VSC	Voltage-Source Converter
VSI	Voltage-Source Inverter
WIP	Work In Process
ZBB	Zinc-Bromide Battery

Foreword

Dear reader,

Welcome to a new world.

With *Green Electrical Energy Storage, Science and Finance for Total Fossil Fuel Substitution*, you have in your hands the most extensive introduction to the new world that energy storage is opening in front of all of us.

Dr. Zini made me the honor to ask me to write the preface to this book, and I am grateful for this. Indeed, this masterpiece is an absolute must-read both for newcomers, who should do their best to slowly digest the entirety of the book, and for professionals not yet fully aware of some of the many technical, legal or financial aspects of the industry.

Energy storage is about to effectively start changing the world of energy as we know it, and as these lines are being written, this revolution is already under way.

I run Clean Horizon, and we have been a consultancy specialized in energy storage ever since I launched the company in 2009. At this time, while the concept of energy storage was clear, its legal ramifications were poorly documented, and the financial community was vastly ignorant of most aspects of the topic (except for Pumped Hydro Stations, which had already been invested in for decades). In the beginning of this decade, we at Clean Horizon were mostly involved in technology watch and in monitoring emerging regulations.

Today, in 2015, multiple MW-level commercial battery projects are being invested in and grid-connected in geographies as varied as the continental US or the French Antilles in the Caribbean. Our work has thus shifted with the industry, with Clean Horizon now being heavily involved in system deployment and business development.

So, in the span of just five years, energy storage has started emerging from RD&D (Research Development and Demonstration) towards effective infrastructure projects with real, bankable, business models. The rapidity and scale of this evolution are probably just the first signs of an accelerating change in the energy storage space. If we were futurologists trying to predict its future, rather that drawing a "straight line" between the data points of the early years of this industry, we would rather draw an accelerating, exponential curve: indeed each new deployment allows more investors and end-users to realize that energy storage technologies can safely function and generate profits, so that each new project carries within it the seeds for more and more deployments to come

So, when we look at the next 5 years of the nascent energy storage industry, it seems clear to us that a "new world" will open up, at increasingly larger scales

For sure, multiple forces are at play to support this new development.

The first one is an increase in demand, mostly driven by higher penetration of renewables creating hurtful intermittencies on electric grids worldwide. This drive, sometimes pushed by green policies but now more and more by economic pressure in the face of cost reductions in solar and wind technology, is probably a long-term trend which will grow stronger in the next 5 years.

On the other hand, offer is also increasing, and this tends to drastically decrease prices for energy storage systems. A striking example of this trend are announcements made earlier this year by Elon Musk, one of the highest profile entrepreneurs of the beginning of the 21st century and the founder of Tesla Motors. Tesla hopes to soon bring batteries from its future "giga factory" not only to its electrical vehicles, but also to the grid and residential energy storage markets.

Whether Mr. Musk ends up executing or not on his vision in the face of strong competition from Asian manufacturers remains an open question. But the fact that the world's highest profile entrepreneurs are now addressing the energy storage challenge should help gauge both the magnitude and the centrality of the role energy storage is about to play for the rest of the decade.

In the face of this burgeoning industry, accessing extensive and quality educational material to learn the fundamentals needed to effectively tackle energy storage is a difficult task. As a matter of fact, energy storage first of all rests on a vast array of scientific and technological knowledge, making use of disciplines as varied as physics, chemistry and thermodynamics. In addition, understanding and computing the profitability of energy storage systems is often still a complex art, relying on a deep understanding of economics and finance as well as of legal aspects.

Therefore, a relevant introduction to energy storage needs to provide solid foundations in these diverse sciences to adequately prepare engineers, financiers and service providers for the rest of the decade and beyond.

With the book you now have in hands, Dr. Zini has achieved this ambitious task to compile in one manual the fundamentals of all the sciences needed to tackle the present and the future of the energy storage industry.

This book is a gateway to this new world, to which the global community of energy storage is glad and proud to welcome you.

So, we wish you a lot of success for this new adventure you are about to embark on. And, above all, we do look forward to hearing from you at an energy storage conference or lecture or on an energy storage project of your own quite soon!

Michael Salomon
CEO
Clean Horizon

Foreword

It was the beginning of 1519 when Hernan Cortes landed with 11 ships on the Yucatan peninsula, leading an expedition that caused the fall of the Aztec Empire and brought large portions of mainland Mexico under the rule of the king of Castile. The first thing he did after setting foot on the shore was to burn the 11 ships to ashes, to make sure that the commitment would be there to overturn the Aztecs.

Fast forward a few hundred years, and after applying some caution to extend history teachings, one should perfectly understand the breadth and purpose of what the author of this book set out to do. Renewable energies have been known and used for years. Some, like hydropower, with larger usage than more recent ones like solar in its various declinations, but all have unfortunately been at the margin of the energy production mix of developed and less developed nations. The main reasons for the marginal nature were higher cost, lack of economies of scale, and, obviously, uncontrollable nature of natural energy sources.

Energy storage has also been known for a long time, easier or more difficult to apply in its various interpretations, from mechanical, like in flywheels and pump storage, to chemical as in batteries and fuel cells, and other technologies just now reaching commercial stage. As in the case of renewable energies, energy storage has played at the margin, for quite similar reasons, although compounded by the immaturity of some of those technologies.

Now, both renewable energies and energy storage have become two of the hottest topics in the institutional and private investing space as both have the potential to disrupt existing systems based on conventional fossil fuels, as well as decades of further growth potential in production of and applications related to fossil fuels. However, both renewable energies and storage have to go hand in hand to deploy their full disruptive potential.

I am not trying here to cast support to what happened back in 1519, but the time is ripe to burn the ships and be fully committed to conquer a cleaner future for the people that will follow us. The technical and commercial tools on how to achieve such end are fully and clearly described by the author of this book.

I am very proud to have had the chance to collaborate and share ideas with Gabriele, a distinguished and unique professional, and a friend, and hope others will have the same chance.

Alberto Dalla Rosa
Partner and CEO
Amplio Energy

Foreword

The development of efficient and viable large-scale storage of electric energy is one of the oldest and most complex challenges that technologists have been facing since the electrification of energy transmission and distribution starting from the second half of the 19th century. Over the decades, scientists and engineers have been successful in tackling the many issues that they were presented with, and many different technologies have been developed with characteristics that make them fit for use in any kind of electric power plant.

Only recently though, the huge penetration of renewable energy conversion plants all around the world as a response to climate change has spurred a novel interest in the deployment of storage plants for a myriad of different intents and applications. Just to name a few, storage systems are used in private houses to optimize the use of solar energy converted by rooftop-mounted PV panels at night. Rural areas use solar energy and batteries for lighting so that young kids can study at night or to increase shelf life of pharmaceuticals; firms use energy storage to compensate for electric grid instability, and utilities to regulate frequency variations caused by imbalance in the demand and supply of electric energy.

In most electric systems, the uncoupling of power generation and utilization provided by energy storage systems can determine remarkable advantages. In several countries, for instance, the environmental impact of electric power generation increases due to the power plants used to cope with the power demand peaks, often characterized by a low efficiency. The integration of energy storage systems can reduce the use of these plants. Moreover, the increasing demand of sustainable and reliable energy sources imposes increasingly stringent regulations on the admissible quality of the power delivery to the electric grid, which can greatly benefit from the integration of storage systems.

For many years, the Department of Electrical, Electronic and Information Engineering of the University of Bologna has offered courses on renewable energy sources and energy storage systems and has started research programs recognizing their strategic importance and impact on the development of future electric power systems. Research activities have been focused both on the development of more efficient energy storage systems and on their optimized implementation in the electric grid or in automotive applications. For development of the components, efforts have been devoted to increase the efficiency of lithium-ion batteries for automotive applications through the introduction of innovative nanofibrous separators. Moreover, activities concerning the optimized design, manufacturing, and testing of Superconducting Magnetic Energy Storage (SMES) systems realized starting from different superconducting materials have been carried out and are presently undertaken in the frame of national projects. Between the various collaborations, our department is also actively working with the author of this book on the modeling and design of novel energy storage systems. From the electric power system standpoint, research activities have been focused at increasing the efficiency and reliability of the energy generation and distribution through the integration of storage systems and renewable energy sources either in the broad utility grid or in small, independent power systems and in modeling the state-of-charge of secondary cells employed in micro-grids that include the presence of distributed generation from renewables.

The wide variety of energy storage systems, ranging from electrochemical batteries, pumped hydroelectric, compressed air energy storage, and flywheels, to supercapacitors, thermal energy storage, SMES, hydrolyzer/fuel cell systems, and other applications makes this topic a highly interdisciplinary one. The body of knowledge needed for work or research in this field is therefore extremely varied and scattered; books like this one consolidate in a single source of information numerous concepts and techniques that can minimize the time spent on the learning curve and constitute a valid reference for future perusal. Most notably, the book covers not only technology but also concepts that relate to, for instance, legal, finance, or risk analysis, that should be understood by scientists and engineers due to their paramount importance for the successful development of energy storage solutions when taken from the academia to the economic world.

Prof. Marco Breschi, Ph.D.
Department of Electrical, Electronic and Information Engineering
Alma Mater Studiorum—University of Bologna

Introduction to Green Electric Energy Storage

Summary. Environmental concerns, together with the diminishing availability of traditional fossil fuels, have spurred the adoption of green energy–conversion technologies and increased their penetration in many countries. Green energy is most often converted to electric energy, since electricity is one of the most usable forms of energy. Unevenness of many forms of renewable energy sources have impacts on the management of electric energy distribution grids. Storage of energy is a solution to better management of distribution grids. Storage of electricity vastly improves the benefits brought about by green energy technologies. This chapter aims at providing a rationale for the use of green energy and a common framework by defining the main concepts used in the industry.

The Risk and the Response

In recent years, numerous scientific analysis have caused a lot of concern about the sustainability of our economical system based on climate-changing and polluting fossil fuels [1–7]. The correlation between mankind activity and climate change has been shown as being so high as to cause an alarmed and widespread response by public opinion and their governments.

From the information publicly available on web sites of widely recognized and renowned organizations [8], it is possible to get the following alarming pieces of information:

- Earth temperature has risen 1.5°C since 1880

- The decade between 2000 and end of 2009 has been the warmest since records have ever been taken

- In 2012, the Arctic summer sea ice has reached its lower extent of −12% over a decade

- Greenland is losing around 100 billion tons of ice every year and the loss has doubled in the decade between 1996 and end of 2005

- The sea level is increasing by more than 3 mm per year and has risen between 10 and 20 cm in one century

All of these changes in the vital signs of our planet Earth are considered as being highly correlated with an extremely high rise in the concentration of carbon dioxide (CO_2) which has reached in 2012 a level of 396 parts per million, the highest in 650,000 years.

Carbon dioxide is a very powerful green-house gas which is released both through human activities (such as combustion reactions and deforestation) and natural processes (respiration and volcanic activity). Data reconstructed from ice cores show that the highest value ever recorded previously in history has a value of nearly 300 parts per million around 325,000 years ago, where only natural processes were occurring. Today, carbon dioxide levels are close to 25% higher than the highest historical level, with this increase all happened during the last century, with human activities considered as the likeliest cause of change.

Around 88% of the energy sources that are adopted in the global economic system is based on fossil fuels. Fossil energy is highly correlated with significant damages to the global ecosystem. The current energy infrastructure is no longer sustainable for our planet and our species. A shift to a different way to organize our energy economy seems to be therefore necessary and inevitable.

Fossilization of vegetation and animals has been a formidable way to clean our atmosphere from elements (mostly carbon, hydrogen, and oxygen) that would have prevented life on our planet to express itself the way it is today. Those elements were captured in the form of hydrocarbons, highly energetic complex chemical structures that have been the major source of energy retrieval for humans during the last decades.

Other energy sources that come from different natural processes are as well widely available in our planet. These energy sources are renewed over very short periods of time or are constantly provided by our ecosystem; the term *renewable energy* (REN) is nowadays commonly used to distinguish them from fossil energy and, since renewable energy sources are also free from environment pollutants, the term *clean energy* or *green energy* is also used in the common language.

Green energy has always been available to mankind: energy from the Sun, from the Earth, and from the gravitational forces acting between our plant and the Moon, has always been known and used, much before the use of hydrocarbons. As a matter of fact, solar radiation can be converted to more usable forms of energy, like electricity or combustible fuels, by technologies such as photovoltaic (PV), concentrated solar power (CSP), wind and wave harvesting, hydroelectricity, biomass direct combustion, or gasification; earth energy is available under the form of geothermal energy; gravitational energy can be exploited by tidal energy conversion.

Many countries in the world have been at the forefront of such growing awareness and many have committed to extremely meaningful programs of shifting from fossil fuels to renewable energy sources in an effort to reduce the impact on our world ecosystem. Fossil fuels play such a significant part on the release of carbon dioxide in the atmosphere that resorting to renewable energy sources is considered as one of the most important and fast ways to cope with the need to reduce CO_2 emissions from human activities.

A number of countries have implemented very aggressive programs to boost the adoption of new energy efficient technologies and the construction of power plants based on the conversion of renewable energy sources. Some of them (like Spain, Italy, Germany, and Denmark) have been extremely successful in increasing the penetration rate of renewable energy sources in their existing energy infrastructure. As a downside, all such countries have experienced unexpected problems caused by the unpredictability of such new sustainable energy sources. Both PV and wind energy change in yield by the minute, by the day, across the the whole year, reducing predictability of energy injection and potentially increasing voltage and frequency instabilities. Therefore, a country's energy infrastructure with a large penetration of intermittent renewable energy sources can become extremely difficult to manage.

The increasing introduction in the grid of renewable energy power plants has not only led to important environmental and economic benefits, but has also placed new challenges on transmission and distribution (T&D) of electricity in power networks. As a result, fossil fuel traditional energy systems cannot be substituted by the renewable energy sources the way they are nowadays implemented.

Energy storage systems (ESSs) are the solution to this problem and, if combined with the conversion of energy from renewable sources, grant the chance to drastically reduce, or cancel altogether, its dependence on the fossil fuel infrastructure. Between the many benefits, big centralized electric storage power plants or distributed installations (i.e., for residential, community, and industrial applications) would facilitate and increase the introduction of large shares of renewable energy plant connected to the grid. If the grid needed increases in energy supply, decentralized ESS could provide part of the capacity by discharging in conjunction with other decentralized storage units or, on the contrary, by absorbing part of the surplus energy available in the grid in case of supply larger than demand. This way, even with varying power injection, the grid would still remain stable and reliable.

Adoption of ESS is highly correlated with the introduction of new smart-grid technologies. Anyway, real-time management is not sufficient to cope with the new challenges tied to the substitution of fossil fuels. Electric energy has to be delivered not only to different locations, but also at times that can be different from the moment that such energy has been produced.[1] ESS play thus a prominent role in uncoupling the moment in which energy is "made" from the moment in which the same energy has been used. Energy storage is then the key to enable decisions about when to use the energy converted from natural sources that cannot be time controlled. ESS applications are therefore considered a prominent medium-term development for energy networks.

Ultimately, energy storage is needed because many trends are already changing the energy infrastructure in many countries around the world. A strong push toward better energy-management technologies, with energy storage being one of the main enablers, is provided by the need of reducing the dependency from foreign energy supplies, by the shift to vehicle electrification, the increasing use of renewable energy sources, the end user propensity to increase energy self-consumption of self-produced energy, and the improvement of grid power quality [9].

The Storage of Electric Energy

Storage can be employed for *power-intensive* or *energy-intensive* applications. If mostly power is needed for relatively short periods of time, the application is power intensive. If energy is needed for relatively long periods of time, the application is energy intensive.

---

[1] In this book we will, sometimes, follow the habit of writing the incorrect term energy "production" instead of "conversion."

ESS are not like conventional power-generation plants; instead of being classified and defined only by their power rating, ESS have both power and energy ratings. This means that the selection of ESS can be fine-tuned to assist for energy-intensive or power-intensive applications and that different storage technologies are not necessarily apt at fulfilling all the range of applications.

Storage can also be intended for *stationary* or *non stationary* uses; stationary installations occur when the ESS supplies energy to a non moving load, while non stationary installations are usually for moving loads, like vehicles, boats, ships, aircraft, material-handling forklifts, and so on. Storage characteristics of stationary and non stationary ESS can be considerably different.

Between the many parameters that define the performance of an energy storage system, the following can be considered as the ones that receive the most of attention.[2]

- *Power*, the nominal power output that an ESS can provide, measured in watts (W)
- *Energy*, the total energy output that an ESS can provide, usually measured in watt-hour (Wh)
- *Capacity*, the charge content in the storage, measured in coulomb (C)
- *Discharge time*, the time period during which the system is able to deliver electricity to the load by conversion of the stored energy of the device
- *Charge time*, the time period during which the system is able to store energy coming from the charging energy source
- *Round-trip efficiency* (RT), indicates the ratio between the amount of energy discharged from the device and the amount of energy necessary to charge it
- *State of charge* (SOC) (minimum and maximum), the minimum or maximum threshold of ESS total capacity that cannot be surpassed in order to avoid damages to the technology
- *Depth of discharge* (DOD), the range between minimum and maximum SOC
- *Life-time* (in cycles or units of time), the number of discharge and charge cycles the energy storage device can experience before it fails to meet specific performance criteria
- *Parasitic losses*, an inner characteristic of some ESS that cause them to lose a quota of their charge, thus reducing performance efficiency
- *Response time*, the length of time it takes the storage device to start providing or storing energy on demand
- *Gravimetric density*, the amount of energy that is stored per unit of weight of the storage system
- *Volumetric density*, the amount of energy that is stored per unit of volume of the storage system
- *Operating parameters*, the environmental or technical parameters that need to be controlled and maintained in order to secure regular ESS functioning

---

[2] Please note that some of them are more relevant to electrochemical storage, as it will become clearer during the reading of this book.

Energy Storage Applications

From final users (residential, communities, or industrial customers) to large-scale utility companies, national or intra national power backbones, all entities connected or even not connected to the grid can benefit from energy storage technologies [10–12]. ESS can effectively complement, or substitute, current power plant technologies according to a set of applications that can be categorized as:

- Electric energy grid services
- Electric power grid infrastructure services
- End user energy management
- Renewable energy management

Electric Energy Grid Services

Electric energy grid services are the set of applications of ESS to services that can be provided by electric energy storage to the T&D electric grids. Such services can be offered either by utilities or by other potential players, different from traditional grid operators, that can seek revenue generation by operating assets to buy and sell electricity in the regulated markets. They can be summarized as the following:

- energy time-shifting
- load following
- up- or down-regulation
- electric supply reserve capacity
- voltage support

Electricity markets are usually structured as day-ahead, intra day, ancillary, and derivative markets. In *day-ahead* and *intra day* markets, electric energy is sold and bought according to auction procedures the day before or during the same day when energy is exchanged between the negotiating parties. The *ancillary* market is where services are purchased by regulatory bodies or grid operators, and such services can be paid for power rather than for energy; an example can be *capacity payment* contracts, where the power from a generating unit is rented and paid even if that generator is not providing energy to the grid, but can be employed upon request when there is a need for its additional generating capacity. Finally, *derivative* markets are based on the structuring of financial instruments (like futures or options) that have energy as the underlying asset.

Energy Time-Shifting

Energy time-shifting entails the arbitraging of energy by buying electric energy after having charged the ESS at off-peak prices, and selling electric energy by discharging the ESS at peak prices. Due to its economic nature, high conversion efficiency and low operating costs for ESS are mandatory to avoid the reduction of the revenues coming from differentials between peak and off-peak prices. If the difference between peak and off-peak prices during the day are not high enough to compensate for conversion losses and operating costs, the ESS logic could decide not to discharge the ESS, therefore reducing its usage ratio and diminishing the returns for investors and owners. In energy markets where RES have a significantly high penetration, this application can cater for increases in revenues by avoiding the sale of energy when renewable energy injection is at its highest but, at the same time, energy market prices are at their lowest.

The use of ESS for energy time-shifting improves the stability of the electric grid; by acting as a balancing system, the ESS can provide a leveling effect avoiding under- or over-utilization of the grid.

Load Following

Load following is the service needed to compensate the variations in grid power needs that occur as frequently as every several minutes due to changes in hourly load curves or in power flows from generation between interconnected grids. Load following is *up* as power output is increased (i.e., with new generation assets that are connected to the grid) to match an increase in demand, or *down* if power output must decrease (i.e., assets disconnected) to reduce power supply in case of diminishing load needs.

Frequency Regulation

Frequency regulation is the service provided for the stabilization of the grid frequency that varies as a result of short-term, continuous differences in demand and supply. If power injection and power absorption to and from the grid are not balanced, the energy network becomes unstable and unreliable. Frequency regulation is similar to load following, but operates on a continuous time scale.

When there is an increase in the load or a decrease in traditional power generation, the other power generators slow down due to load increase. As a consequence, overall grid frequency tends to decrease. If this situation is not counteracted, grid frequency would decrease up to the point when the connected loads would no more be capable of working at their lower-bound operating conditions. Power units connected to the grid must therefore increase their power output to supply the additional load; this is called *up-regulation*; when power units must decrease their power injection due to an increase of power supply, this is called *down-regulation*.

For the effectiveness of load following, frequency regulation, a coordination system between power units in an electric power system is required. An *automatic generation control* (AGC) system adjusts the power output of all units at different power plants by taking as a reference parameter the grid frequency: a drop in frequency is seen as a signal to increase power output, while an increase in frequency as a signal to reduce power output. For instance, in traditional turbine power plants, *turbine governor* controls automatically maintain the desired system frequency by adjusting the mechanical power output of the turbine.[3] Another way to coordinate area-wide power increases or reductions is the use of *Area Control Error* (ACE), an algorithm that computes the difference between budgeted and actual power injection in the grid using frequency as the real-time input parameter [15].

---

[3] One common way to control system frequency is using *droop speed control*, which uses net frequency deviations to distribute load over power plants. For instance, power plants can be made to operate with a 5% speed droop, meaning that the power unit full-load speed is 100% with a no-load speed of 105%, used by the governor control unit as a power slack to compensate for frequency variations.

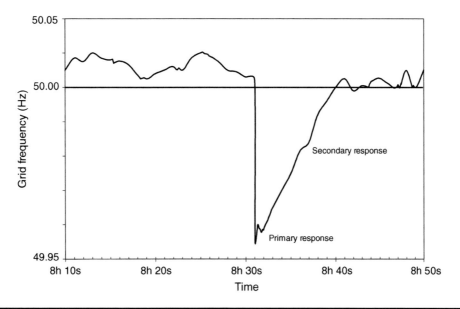

FIGURE 1 Events following a contingency over a power grid.

Grid frequency value is constantly monitored by the grid operator to trigger two kinds of response (see Fig. 1).

- *Primary frequency control*, also called *primary response*, is performed by the power units governors and occurs in the first seconds of decrease in the grid frequency (the *arresting period*). As soon as the governors start the response, the frequency begins to increase, with a *rebound period* which occurs usually after 10 s from arrest and lasts as long as 15–20 s

- *Secondary frequency control*, also called *secondary response*; after the rebound period, the power units adapt their output through AGC up to 10 min or more after the arresting period (the *recovery period*)

Electric Supply Reserve Capacity

The *electric supply reserve capacity* is the reserve capacity that can be made operational in the short term to compensate when part of the normal power supply becomes unavailable. it consists of power generation units that are normally disconnected from the grid but can be connected quickly when there is an unexpected surge in demand or after the inception of technical problems (i.e., shortfalls and outages). The service is also known as *short-term operating reserve* (STOR), and entails three levels of intervention.

- *Spinning reserve* is the generation capacity that is ready to be loaded within 10 s from notice. Spinning reserve is maintained synchronized with the grid frequency. The reserve can even be made available by generators that are already connected to the grid and are not at full capacity. Also called *primary reserve*

- *Non-spinning reserve* is power capacity which is not synchronized and not connected to the grid but can be loaded within a short period of time, usually in the range of 10 min. In interconnected grids, this capacity can be obtained not only by the use of fast-start generators but also by redirecting power from other grids. Also called *secondary reserve*

- *Supplemental reserve* is a back-up capacity that can be made available to the grid in time-frames of 30 min to 1 h and is used after spinning and non spinning reserve facilities have already been used. Supplemental reserve is also referred to as *replacement reserve*, *contingency reserve*, or *tertiary reserve*

Typically in electric grids, the reserve capacity is 15–20% of the total capacity normally available. Frequency regulation can be considered within the STOR services as *regulating reserve* (also referred to as *area regulation*).

There exist also alternative ways to address capacity issues occurring in the grid. For instance, utilities can curtail the power needs by disconnecting loads with or without notice to avoid larger outages in the whole of the T&D grid. An example of agreed power curtailment are the *interruptibility* contracts, which are agreements entered into by a transmission/distribution grid operator and a final user (i.e., an industrial customer) who accepts to be interrupted on an instantaneous basis (with response times that can be in the range of 200 ms) or after being notified in advance, against the payment of a per-interrupt, monthly, or annual fee.

Load following, frequency regulation, and STOR are difficult to be performed by large thermoelectric power units due to operating costs and reduced asset life time. ESS on the contrary, are particularly suited for such applications due to their fast response times and quick switching between charging (downregulation) and discharging (upregulation) modes [14]. A 10 MW ESS can therefore operate as a 20 MW STOR unit. For these services, the size of the ESS and the position in the grid are important parameters to be carefully assessed in order to maximize quality of intervention.

Voltage Support
Voltage Support is the stabilization of the grid voltage level by compensating for the changes in the reactance caused by the different combination of inductive or capacitive loads connected to the grid. ESS can act as devices that absorb or inject reactive power. ESS are indeed capable of operating either in charging or discharging mode while at the same time providing VAR capability or working in VAR-only mode without injecting or absorbing to and from the grid active power but only reactive power.

Table 1 summarizes the applications needed to assist electric energy grid services.

Electric Power Grid Infrastructure Services
Electric power grid infrastructure services concern the use of electric energy storage by utilities to support and maintain their T&D power line operations. Such applications comprise of the following:

- T&D support
- T&D congestion relief
- T&D upgrade avoidance/deferral
- substation on-site power

Application	Energy time-shift
	Purchase of electric energy by charging off-peak, sale of electric energy by discharging on peak
Capacity	1–2 h
Power range	1–500 MW
Cycles/year	365–730
Notes	Need for high round trip efficiency, low operating costs
Other terminology	Arbitrage
Application	Load following
	Balancing of grid power supply and demand needs by increasing or decreasing power output
Capacity	30 min to 2 h
Power range	10–100 MW
Cycles/year	Depending on circumstances
Notes	ESS can provide two times its capacity as reserve capacity (by charging and discharging)
Other terminology	Fast reserve
Application	Frequency regulation
	Continuous balancing of grid power supply and demand needs by increasing or decreasing power output
Capacity	10 min to 1 h
Power range	10–50 MW
Cycles/year	Full discharge per cycle is not necessarily reached, but cycling can be frequent and much higher than 200 cycles per year
Notes	ESS can provide two times its capacity as reserve capacity (by charging and discharging)
Other terminology	Regulating reserve, area regulation
Application	Electric supply reserve capacity
	Reserve capacity operational in the short term to compensate when part of the normal power supply becomes unavailable
Capacity	1–6 h
Power range	10–500 MW
Cycles/year	50–365
Notes	ESS can provide two times its capacity as reserve capacity (by charging and discharging). Response times in the range of seconds or 1 h depending on application
Other terminology	Short-term operating reserve (STOR)
Application	Voltage support (VAR)
	Injection or absorption of reactive power to the grid
Capacity	30 min
Power range	1–10 MVAR
Cycles/year	No discharge taking place, number of cycles losing significance
Notes	Short or very short response times (ms)
Other terminology	

TABLE 1   Outline for Electric Energy Grid Services Applications

T&D Support

T&D support is a set of services to maintain optimal functioning of the transmission grid. One typical service is the *Black Start* of transmission lines or power components that can be provided by ESS to reenergize transmission lines and transformers after grid maintenance stops or crashes. Other services can entail providing active control for the improvement of the dynamic stability of the grid. Since power flow in transmission lines is a function of line impedance, certain power systems can experience frequency-deviation effects that can cause instability phenomena. For example, *Sub-Synchronous Resonance* (SSR) is a frequency-deviating effect caused by the resonance between turbine shafts and the grid series capacitors needed to filter out part of the power line inductive reactance, resulting in a lower than nominal line frequency [16]. As a result, resonance between mechanics of the turbine and the electrics of the power system can produce torsional stresses on turbines resulting in potential severe damages to the generators eventually leading to breakages, failures and grid outages. Transmission stability applications usually require power-intensive rather than energy-intensive ESS, while black-start ESS need to have large capacities to recharge the grid and power components.

T&D Congestion Relief

T&D congestion relief applications aim at avoiding the congestion-related costs when transmission lines are subject to power overdrawn. This service shaves the peak above a pre-determined power threshold by discharging a locally installed ESS that is charged when the transmission line is less congested.

T&D Upgrade Avoidance/Deferral

T&D upgrade avoidance/deferral is the possibility, granted by ESS, to delay or avoid the grid system upgrade when the connection of additional loads nears the grid maximum load-carrying capacity as per initial system design. In case peak loads were reached only for a few hours during the week or even months, installing ESS to serve those incremental loads could avoid investments in reinforcement of power lines or new equipment that would have very low load factors during the year and result therefore unprofitable to stakeholders. ESS would add the increased capacity directly close to the loads to be served, creating a more decentralized electric power system with additional benefits on the system reliability and quality of service. As a further additional benefit of this application, the usable life of old equipment like transformers or underground cables can be extended since the use of ESS can reduce their load. Expected end-of-use can therefore be postponed farther away in the future thanks to less aging from decreased load factors.

Substation On-Site Power

Substation on-site power applications entail the use of ESS for stand-alone, isolated energy supply systems for telecom stations, control sites, power switching components, desalination plants, homes, that would have to be connected to a distribution line but at costs that would make construction unfeasible. As a result, use of ESS makes electrification in rural areas possible.

Table 2 shows an outline of ESS applications for electric power grid systems.

Application	T&D support
	Response to electric instabilities, energization of lines, and components after stops
Capacity	Seconds to 2 h for stability. 15 min to 1 h for black start.
Power range	1–100 MW
Cycles/year	Above 15
Notes	Very quick response times (milliseconds). Difficult to be used concurrently with other applications
Other terminology	
Application	T&D congestion relief
	Avoidance of congestion related costs and charges due to transmission system congestion.
Capacity	1–4 h
Power range	1–100 MW
Cycles/year	Above 50
Notes	
Other terminology	
Application	T&D upgrade avoidance/deferral
	Delaying or avoidance of investments in T&D system upgrades or extension of useful life of T&D systems
Capacity	2–6 h
Power Range	5–100 MW
Cycles/Year	Above 10
Notes	
Other Terminology	
Application	Substation on-site power
	Power to switching components, telecom substations, control units, rural equipment.
Capacity	8–16 h
Power range	100 kW to 1 MW
Cycles/year	Up to 365
Notes	
Other terminology	

TABLE 2   Outline for Electric Power Grid System Applications

End-User Energy Management

End-user energy management is the set of services that ESS can provide to final users like residential, community or industrial customers. Such services are the following:

- time-of-use cost management
- demand charge management (peak shaving)
- power quality and reliability

Time-of-Use Cost Management

Time-of-use cost management (TOU), also called *load leveling* or *energy shifting*, involves charging of ESS during off or partial peak hours and its discharging during peak hours to save on electricity costs. Knowledge of peak, partial and off-peak prices, and time ranges is straightforward for small customers and ESS programming is easily feasible.

Demand Charge Management

Demand charge management, very often referred to as *peak shaving*, is the provision of electric power to save on the electricity price apportioned to supplied total power. If the ESS is discharged when power draw overcomes a predetermined threshold, the power is supplied by the ESS rather than the grid therefore reducing the maximum power supplied to the end user by the grid operator. Usually, power draw is checked over a 15-min period and billed monthly. Shaving the 15 min highest monthly power draws above a certain threshold can result in substantial savings. This strategy can be unachievable if the load profile is flat since the battery would have to be sized with a very high capacity only to get power draw price reductions that are normally marginal if compared to the sale of energy by, for instance, arbitraging.

Power Quality and Reliability

Power quality and reliability services provide fast response power and energy injection to maintain high standards of the quality of service (QOS) in the customer grid. In case of long outages, ESS can be used to provide uninterrupted service (like emergency back up) for residential customers or maintain stable production output for industrial customers or provide enhanced safety and damage avoidance by safe machine shut downs.

Table 3 provides a summary of ESS applications for end-user energy management services.

Renewable Energy Management

Renewable energy sources have synergies with ESS that increase their value for the overall energy system and reduce the need to rely on traditional fossil fuel energy generation. Two applications can be devised:

- renewable energy injection profiling
- renewable energy injection smoothing

Renewable Energy Injection Profiling

Renewable energy injection profiling, also called *time-shifting*, makes it possible to choose when energy from renewable sources can be injected in the grid. This can be useful to improve renewable energy asset profitability as well as controlling the energy flow to the grid to avoid injection when demand is lower than supply.

Renewable Energy Injection Smoothing

Renewable energy injection smoothing consists in filtering the uneven energy injection profiles typical of wind or solar PV conversion. ESS can be charged when RES energy is above a certain threshold and discharged to smooth peaks and troughs to avoid grid instability. ESS are one of the most interesting solution to this issue and can foster larger quotas of RES penetration in the electric power system. ESS need to have high round-trip efficiency to avoid energy losses and maintain productivity and profitability of RE assets.

Table 4 summarizes the uses of ESS with RES.

Application	TOU cost management
	Reduction of electricity costs by postponing the use of stored energy from off-peak to peak hours
Capacity	1–5 h
Power range	1 kW to 10 MW
Cycles/year	Up to 365
Notes	
Other terminology	Load leveling, load shifting
Application	Demand charge management
	Power supply from ESS over a certain power threshold to reduce power draw pricing
Capacity	15 min to 4 h
Power range	1 kW to 10 MW
Cycles/year	50–500
Notes	Need for fast ESS response times
Other terminology	Peak shaving
Application	Power quality and reliability
	Use of ESS to provide high QOS and reliable operations.
Capacity	Hundreds of milliseconds to 15 min
Power range	1 kW to 10 MW
Cycles/year	10–200
Notes	Need for fast ESS response times
Other terminology	

TABLE 3 Outline for End-User Energy Management Applications

Application	Renewable energy injection profiling
	Time-shifting of RE energy grid injection
Capacity	1–12 h
Power range	1 kW to hundreds of megawatts
Cycles/year	365
Notes	Need for high ESS RT efficiency
Other terminology	Energy time-shift
Application	Renewable energy injection smoothing
	Smoothing of peaks and troughs in RES energy conversion profiles
Capacity	1–12 h
Power range	1 kW to hundreds of megawatts
Cycles/year	Above 300
Notes	Need for high ESS RT efficiency
Other terminology	RE firming

TABLE 4 Outline for Renewable Energy Management Applications

Use of Energy Storage Systems as a Substitute to Fossil Fuel Power Plants

ESSs, if combined with either traditional energy or innovative clean energy production sources, can partially or totally substitute for traditional technologies according to studies and reports from respected public-funded or private research international institutions (see the reference section of this chapter).

The power plants that are connected to the electricity grid must provide basically three types of power services. *Base-load* power is the power that is always requested by the loads connected to the grid, and is normally around 30–40% of the highest power draw from the grid during a day. Coal-fired, nuclear, hydroelectric, or biomass are the technologies used as base-load power plants. Base-load power plants can take many hours or days to change their power output. The economics of their operations rely on low operating costs, thanks to energy production programs that are planned in advance. Planning facilitates optimized fuel purchase long-term contracts bringing about reductions in costs, and makes possible operating the power plants at their highest efficiency. *Load-following* power plants change production in order to match the varying power consumption demand above the base-load of the area they serve. Normally, natural gas steam turbine or hydroelectric power plants are used since they can provide fast changes in their power output. *Peaking* power plants inject power to the grid only in cases when there is an occasionally high demand for energy. Normally, natural gas steam turbine plants are used as peakers, often in a highly efficient combined cycle gas turbine configuration.

By combining ESS to a high penetration of RES in the power generation mix and a new paradigm in energy management smart technologies, base-loading can be covered by technologies that do not resort to fossil fuel usage; entire regions with varying energy availability and demand needs can be interconnected with smart power backbones that can make energy flow consistent with supply-demand logic.

For instance, replacing a 1 GW coal-fired power plant would need a cluster of 2600 MW wind turbine, where higher capacity caters for the same base-loading capabilities of the coal-fired plant. It would seem that the main issue could be land occupancy, but better scrutiny would yield that the surface occupied by the wind energy turbine cluster could cover an equivalent area of 10–20 km^2 depending on wind speed in the area, much less than the surface needed by the coal-fired plant and its associated open-air coal mine that occupies an area close to 50 km^2. Same reasoning could be applied to solar PV, while CSP[4] plants with 7–8 or more hours of storage capacity can already represent a viable solutions for base-load fossil fuel substitution [17].

Even power-peaking can be managed more effectively by ESS coupled with renewable energy sources. A study by the California Energy Storage Association has shown that by using 100 MW of ESS instead of a 100 MW gas turbine peaker plant, the benefits that could be gained would be the following [18]:

- 600 times faster ramp up of power supply
- 4 times the flexibility of its range of use

---

[4] A concentrated solar power plant converts solar radiation to thermal energy which is stored in molten salt reservoirs and then converted back to electricity in turbines using Brayton's or organic Rankine's cycles.

- 3 times more available operating hours
- 6500 gallons/h of water usage saved
- 90% GHG emission reduction

A gas peaker would indeed have a ramp up time of 10 min instead of less than 1 s for an ESS, 50 MW of flexible power range instead of more than 90 MW for an ESS, 2768 usable hours per year instead of more than 8300 h for an ESS.

As a conclusion, base-loading and peaking plants, with load-following based on a mix of the two, can easily be achieved by renewable energy sources combined with electricity storage.

Benefits and Challenges of Energy Storage Systems

As discussed in the previous sections, many are the benefits that ESS provide to the energy sector.

ESS facilitate increase of penetration of renewable energy power plants by stabilizing their energy injection in the grid. ESS combined with RES can be valid substitute for traditional fossil fuel–based power plants. As large penetration of uneven renewable energy sources are affecting grid stability, avoiding further introduction of RES or curtailing their power injection is not the solution, rather, grid stabilization is the key to increase the share of energy converted from renewable sources.

ESS provide increased efficiency, multi-purpose, and larger utilization rates than actual technologies, thus greatly reducing fossil fuel usage and infrastructure costs.

ESS are inherently distributable and scalable. Grid efficient use and better management of existing T&D line can be achieved with the use of ESS, with further reductions in greenhouse gas emissions. Cost reductions are achievable since adding ESS after a grid bottleneck and close to the load can postpone, or avoid altogether, the installation of additional cables in parallel to the existing ones.

The grid can be rendered safer and more reliable, since overloads can be avoided and the frequency regulated by using storage to supply power during peak demand and absorbing when in oversupply. The grid will also be more resistant to outages due to bad weather conditions or other natural or man-based threats.

Finally, ESS are also faster to be commissioned than traditional generation plants, improving the quality of the infrastructure by adding power assets in shorter times than traditional fossil fuel plants.

Some of the benefits that can be captured by the use of ESS can be summarized as:

- power backbone stability
- increased introduction of renewable and clean energy sources
- reduced fossil fuel usage
- efficient use of power grid infrastructure
- scalable and flexible capacity
- fast commissioning of new energy infrastructures
- decentralization of power supply
- energy supply independence

Efficiency	> 80%
Reliability	> 98%
Life-cycles	> 5000
Capital expenses (CapEx)	< 200 USD/kWh
Operating expenses (OpEx)	< 1.5% of CapEx
Levelized costs of energy (LCOE)	< 10–15 cents USD/kWh/cycle

TABLE 5 Minimum Requirements for a Successful Deployment of ESS

The challenges for an effective development and successful market penetration of ESS are, at the moment:

- *Efficiency.* ESS should cause the least possible conversion losses during charge and discharge cycles. RT efficiency should remain over 80%
- *Reliability.* ESS should be in operations for the longest possible time between downtimes due to failures or maintenance. A good availability would be over 98–99% of the time
- *Durability.* Decays in performance should be negligible and ESS should last for a long number of charge-discharge cycles. A minimum number of acceptable life-cycles would be over 5000
- *Safety.* ESS must have a track record of safe operations. Pollutants must be confined and non toxic operations must be ensured. ESS must not represent a risk to humans and the environment during commissioning, operations, and decommissioning
- *Cost effectiveness.* Investments in ESS must make business and financial sense. Subsidies should not be considered except for the purposes of starting up a market in an effort to achieve economies of scale, industrialization, and cost reductions
- *Manufacturability.* Large-scale production and lean manufacturing processes will be of the utmost importance to ensure availability, quality, and cost-effectiveness of ESS
- *Deployability.* ESS solutions should be easy to develop and deploy for all potential applications
- *Acceptance.* ESS must represent value in the eyes of the final users; the fact that adoption of ESS can increase penetration of renewable energy power plants, increases acceptance by large shares of the public, and can create a green look to storage technologies

Some of the minimum requirements for feasible mid-term to long-term market acceptance of ESS technology are summarized in Table 5.

Energy Storage as a Value Proposition

An electric power grid comprises of generation plants, transmission lines and distribution lines. Transmission lines are the main backbones of the power grid where electric energy is conveyed with high voltage (HV) cables, while distribution is the dispatching of energy to the final users and occurs in medium voltage (MV) or low voltage (LV) cables.

ESS can add a lot of value to the T&D grid. Fig. 2 summarizes the different applications of ESS as integral parts of the electric energy network, from large power plants to residential homes; all of those market players can find the right energy storage application.

ESS can be used for different applications at the same time. A list of potential synergies between the applications listed in the previous sections are outlined in Table 6.

The Current Status of Energy Storage Systems

Among the currently available energy storage technologies, pumped-hydro counted in 2010 for 99% of the worldwide installed capacity, followed by compressed air storage. The ESS per market share in 2010 were:

1. Hydroelectric: 127 GW_{el}
2. Compressed air storage: 440 MW_{el}
3. Sodium-sulfur batteries: 316 MW_{el}
4. Lead-acid batteries: around 35 MW_{el}
5. Nickel-cadmium batteries: 27 MW_{el}
6. Flywheels: less than 25 MW_{el}
7. Lithium-ion batteries: around 20 MW_{el}
8. Redox-flow batteries: less than 3 MW_{el}

Hydroelectric plants are the most widely diffused form of renewable energy storage so far. Reservoirs are created by damming or channeling water streams to convert water kinetic energy in a turbine, or pumped upstream to replenish water levels in the reservoirs during off-peak hours to take advantage of higher prices during peak hours.

Sodium-sulfur, lead-acid, nickel-cadmium, and lithium-ion batteries are electrochemical storage devices capable of converting electric energy into chemical energy during the charging cycle and converting chemical energy to electricity during the discharging cycle.

The technology of compressed air energy storage has been in development since decades, but not many plants are operational ever since. The earlier systems required a complex underground compressed air storage, but recent technological advances have made above-ground systems a viable option.

Spinning devices like flywheels to store kinetic energy have been used by mankind since centuries; a spoked disk or cylinder capable of storing a significant amount of rotational energy, when shafted to an electric motor/generator a flywheel can act as storage system by varying its angular speed.

Flow batteries consist of a stack of cells where the reactions between two electrolytes take place, with electrolyte tanks and pumps to flow the electrolytes in the battery stacks; some examples of common types of redox-flow batteries are zinc-bromine batteries and vanadium-redox batteries.

A very useful way to categorise energy storage technologies is by means of the *Ragone plot*, providing the status of energy storage technologies according to their energy and power densities.

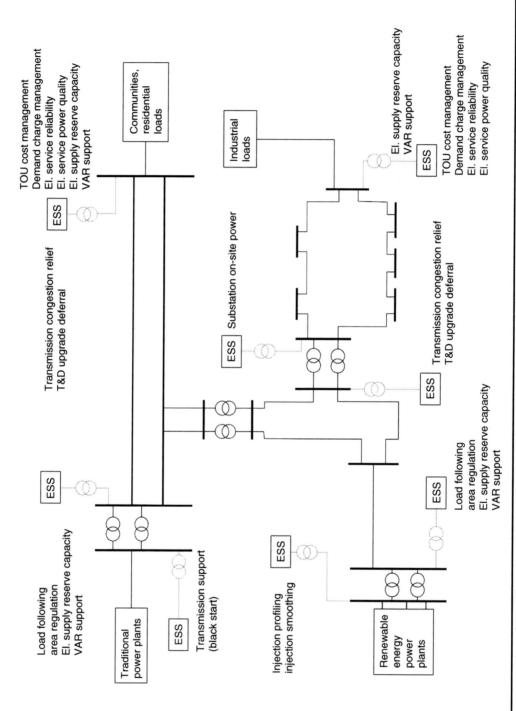

FIGURE 2 Outline of the different uses of ESS over the T&D grid.

	Application	Potential synergies
1	Energy time-shift	2, 3, 4, 5, 7, 8, 12, 13.
2	Load following	1, 4, 5.
3	Frequency regulation	Not simultaneously with 1, 4, 8.
4	Electric supply reserve capacity	Compatible, not simultaneously, with all applications.
5	Voltage support (VAR)	Centralized: 1, 3, 4, 14. Distributed: circumstance specific.
6	T&D support	Difficult to be used concurrently for other applications. Nevertheless, storage used for transmission support during peak demand or peak congestion times could be used at other times for several other applications.
7	T&D congestion relief	All.
8	T&D upgrade avoidance/deferral	1, 4, 10, 11, 12, 14.
9	Substation on-site power	None since stand-alone.
10	TOU cost management	8, 11, 13.
11	Demand charge management	1, 2, 3, 4, 5, 7, 8, 10, 12, 13, 14.
12	Power quality and reliability	None (short discharge duration). If duration extended: all except 3 and 6.
13	RE injection profiling	Centralized; 4, 12. Distributed: 2, 4, 6, 9, 10.
14	RE injection smoothing	Centralized: 3, 14.

TABLE 6 Synergies of Energy Storage Applications

Some Hints on the Future Status of Energy Storage Systems

Evaluations on the market potential for energy storage from prominent consultancy firms [20, 21] show that the prices will drop by a large amount in the years until 2020. Since costs are one of the main issues in the widespread adoption of storage technologies, cost reductions are one of the most sought after achievements in the industry.

Prices, namely the cost plus the manufacturer's margin, will fall due to manufacturing increases in productivity, associated with economies of scale from larger market introduction, spreading fixed costs on higher product volumes and reducing variable costs with improved processes and lower component prices. Some ESS technologies do not show large potential for price decreases, while some other technologies show more promising prospects. Together with scale economies and research and development, the main drivers that are pushing to the development of a large ESS market are the introduction of electric vehicles, research in smart grid devices, and the global increase in the number of grid-connected REN plants. Especially the adoption of electric vehicles with a need to recharge during the night will change the load profile on the distribution grid with ESS becoming indispensable to assist the shift in the load curve.

Another important cost driver is energy storage capacity per volume or weight. Advances in design will increase volumetric or gravimetric densities, therefore reducing costs per kWh stored. Some technological innovations will be captured in the development of cathode-electrolyte pairs that will be capable of increasing cell voltage (i.e., from 3.6 to 4.2 V, cell capacity increase around 17%), new electrodes structures (i.e., layered-layered) to reduce unavailable cell areas (i.e., obtaining capacity increases by 40%), new material, and novel ideas.

In case of a decisive move toward decarbonization, the world community should add an estimated 310 GW of grid-connected energy storage, mostly in countries like the United States, China, Europe, and India. The existing energy policy framework is not apt, in many countries, for creating the best conditions for energy storage deployment; significant distortions in the pricing of electricity are evident in many economies, where the cost of energy is not often transparent. New pricing paradigms must be introduced in energy markets to provide remuneration to the many services that energy storage can provide to the grid or the to end customers.

As an example of the new framework structures that are rising in the most organized and forward-looking countries, the *US Federal Energy Regulatory Commission* (FERC) has issued a number of orders to allow energy storage technologies to enter electricity markets as generation technology (order 719), has recognized the value of fast responding energy technologies like batteries and flywheels for frequency regulation (order 755), has recognized the need to provide pay-for-performance to technologies that add speed, accuracy, resiliency to the energy infrastructures, opening de facto the market for ESS participation.

References

1. Cavallo A. J. (2004). Hubbert's petroleum production model: an evaluation and implications for world oil production forecasts. Natural Resources Research, International Association for Mathematical Geology 4(13):211–221.
2. Hafele W. (1981). Energy in a Finite World: a Global Systems Analysis. Ballinger: Cambridge, Massachusetts.
3. Intergovernmental Panel on Climate Change (2001). Climate change 2001: mitigation. Cambridge University Press: Cambridge, United Kingdom.
4. Lund H. (2007). Renewable energy strategies for sustainable development. Energy 6 (32):912–919.
5. Lund H., Mathiesen B. V. (2009). Energy system analysis of 100% renewable energy systems—The case of Denmark in years 2030 and 2050. Energy 34(5):524–531.
6. Scheer H. (1999). Energieautonomie—Eine Neue Politik für Erneuerbare Energien. Verlag Antje Kunstmann GmbH: München.
7. Moriarty P., Honnery D. (2005). Can renewable energy avert global climate change? Proceedings of the 17th International Clean Air & Environment Conference, Hobarth, Australia. Australian and New Zealand Clean Air Society Publishing, NSW Australia.
8. http://climate.nasa.gov/
9. IEA (2014). Technology Roadmap Energy Storage.
10. U.S. Department of Energy (2013). Grid Energy Storage.
11. Rastler D. (2010). Electricity energy storage technology options a white paper primer on applications, costs and benefits. Electric Power Research Institute (EPRI), 1020676.
12. Abbas A. Akhil, Georgianne Huff, Aileen B. Currier, Benjamin C. Kaun, Dan M. Rastler, Stella Bingqing Chen, Andrew L. Cotter et al. (2013). DOE/EPRI 2013 Electricity Storage Handbook in Collaboration with NRECA, SANDIA Report, SAND2013-5131.
13. Oak Ridge National Laboratories. Spinning Reserve From Responsive Loads. ORNL/TM-2003/19, March 2003.

14. California Energy Storage Alliance (2013). Energy Storage—a Cheaper, Faster and Cleaner Alternative to Conventional Frequency Regulation. Berkeley, California, USA.

15. Glover J. D., Sarma M. S., Overbye T. (2008). Power System Analysis and Design. Thomson, Toronto, Canada.

16. Anderson P. M., Agrawal B. L., Van Ness J. L. (1999). Subsynchronous Resonance in Power Systems. Wiley.

17. Diesenberg M. (2010). The base load fallacy. Energy Science Coalition (http://www.energyscience.org.au).

18. Lin J. (2014). Energy storage applications and benefits. California Energy Storage Alliance. International Conference and Exhibition for the Storage of Renewable Energies, Düsseldorf, 25–27 March.

19. Michael Specht, Frank Baumgart, Bastian Feigl, Volkmar Frick, Bernd Stürmer, Ulrich Zuberbühler (2011). Storing renewable energy in the natural gas grid methane via power-to-gas (P2G): a renewable fuel for mobility.

20. Hensley R., Newman J., Rogers M. (2012). Battery technology charges ahead. McKinsey Quarterly, July.

21. James Manyika, Michael Chui, Jacques Bughin, Richard Dobbs, Peter Bisson, and Alex Marrs (2013). Disruptive technologies: advances that will transform life, business, and the global economy. McKinsey Global Institute.

Green Electrical Energy Storage

PART I

Making It: Science and Technology of Energy Storage

CHAPTER 1
Electrochemical Storage

Summary. Storage of electricity was born with the invention of the first accumulators by the master scientist of the past Alessandro Volta. Since then, the technology has advanced and developed in many different types of batteries using several different types of chemistry. This chapter gives the essential information on the general concepts of electrochemistry, the modeling for the understanding of batteries' behavior and their dynamic simulation, to the different types of technologies currently on the market or ready for commercialization in a few years.

1.1 Introduction

Electrochemistry studies the conversion of electrical energy into chemical energy. One way for this conversion to occur is through *reduction–oxidation reactions* (redox), in which:

- *oxidation* is the loss of electrons with an increase in oxidation state by a molecule, atom, or ion

- *reduction* is the gain of electrons with a decrease in oxidation state by a molecule, atom, or ion

IUPAC[1] defines the oxidation state as the charge of an atom when electrons are counted according to the following set of rules [1] :

1. the oxidation state of a free element (uncombined element) is zero;

2. for a simple (monoatomic) ion, the oxidation state is equal to the net charge on the ion;

3. hydrogen has an oxidation state of 1 and oxygen has an oxidation state of -2 when they are present in most compounds with the exceptions that hydrogen has an oxidation state of -1 in hydrides of active metals (e.g., LiH) and oxygen has an oxidation state of -1 in peroxides (e.g., H_2O_2);

4. the algebraic sum of oxidation states of all atoms in a neutral molecule must be zero, while in ions, the algebraic sum of the oxidation states of the constituent atoms must be equal to the charge on the ion.

---

[1] International Union of Pure and Applied Chemistry, http://www.iupac.org/.

Redox reactions can be spontaneous, with energy provided by the reaction to the environment, or nonspontaneous, where energy is to be supplied to the reaction. Part of the energy that is transferred between the chemical species and the environment can be retrieved or supplied in the form of electric energy, if the reactions take place in systems called *electrochemical cells*.

A metal Me that is immersed in a solution of one of his salts[2] has a tendency to enter in the solution by dissociating in ions Me^{n+} and leaving the electrons $n\,e^-$ in the metal which becomes electronegatively charged. The solution that receives the positive Me^{n+} ions becomes electropositively charged. Such dissociation reaches an equilibrium as per the following:

$$Me \leftrightarrow Me^{n+} + n\,e^- \tag{1.1}$$

where the relative concentrations depend upon the species in the reaction and the environmental conditions.

The metal-solution element constitutes the *electrochemical semicell*. The metal in the electrochemical semicell is the *electrode*, which is the electrical conductor that is in contact with the nonmetallic part of the cell. The nonmetallic part of the electrochemical cell is the *electrolyte*, which is the substance that ionizes the electrode when dissolved in the *solvent* (i.e., water). Electrolytes include most soluble salts, acids, and bases.

As a result:

- the metal electrode loses electrons, is oxidized, and becomes positively charged
- the solution acquires electrons, is reduced, and becomes negatively charged

Depending on ionic concentration, type of metal, and electrolyte, an equilibrium is reached according to Eq. (1.1). As a consequence of the equilibrium, an electrical potential (the *electrode potential*) builds up at the interface between the electrode and the solution. The electric potential value depends on the nature of the equilibrium shift in Eq. (1.1).

If two different electrochemical semicells from two different suitable metals and electrolyte solutions have the two electrolyte solutions connected by an interface permeable to the electrolyte ions only, an *electrochemical cell* (or *galvanic cell*) is formed: an electric potential is present at the two metal electrodes due to the different electric potentials of the two semicells, and the voltage potential has a value given by the difference between the most electronegative electrode potential and the least electronegative electrode potential.

For example, zync and copper can be used in a Zn–Cu cell[3] as shown in Fig. 1.1. The two metal electrodes are immersed in an aqueous solution of $ZnSO_4$ for the zinc electrode and $CuSO_4$ for the copper electrode. When the electrodes are connected by an electron-conductive material, a flow of electrons runs from one electrode to the other, and energy is released to the load connected in series with the electrodes.

---

[2] Following Arrhenius' definition, an *acid* is a substance that produces hydrogen ions, or protons (H^+), in aqueous solutions. A *base* is a substance that produces hydroxide (OH^-) in aqueous solutions. A *salt* is a substance that results from the neutralization reaction of an acid with a base.

[3] This electrochemical cell was invented by J. F. Daniell in 1820, and is known as the Daniell battery.

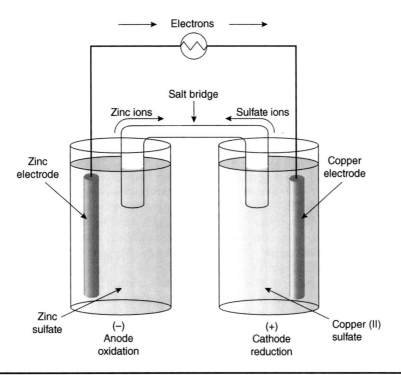

FIGURE 1.1   Schematic of a Zn–Cu electrochemical cell.

The reactions in the Zn–Cu cell are the following:

$$Zn \rightarrow Zn^{2+} + 2e^- \text{ (oxidation)} \tag{1.2}$$

$$Cu^{2+} + 2e^- \rightarrow Cu \text{ (reduction)} \tag{1.3}$$

Zn is the *anode* of the cell, the negative electrode of the cell, while Cu is the *cathode*, the positive electrode. The two reactions (1.2) and (1.3) occur until the equilibrium is reached:

$$Zn + Cu^{2+} \leftrightarrow Zn^{2+} + Cu \tag{1.4}$$

If work is applied to the electrodes in the form of an electric potential with the sign reversed, the reactions occur in the opposite directions and the electrochemical cell is charged, ready to be discharged as soon as such electric potential is disconnected and a load is connected to the two electrodes.

The interface between the electrolyte permits the flow of ions in the solutions to maintain *electroneutrality* of the system and closes the circuit allowing the flow of the electric charges.

To measure the voltage potentials for each electrode material, a *reference electrode* is needed and its potential arbitrarily set to 0 V.[4]

The *standard hydrogen electrode* (SHE) is the *reference electrode* defined and employed to estimate the potentials of each electrode material and chemical element. The SHE is made of platinum around which flows hydrogen at a pressure of 1.01325 bar in a 1 M aqueous solution of $H^{-3}O^+$, and its potential arbitrarily is set to zero.

By connecting to this semicell a semicell from another chemical species, it is thus possible to measure the *standard potential* $E°$ of each element. The standard potentials for all elements are given in reference tables: for comparison, Li^+ has a $E°$ of -3.04 V, Zn^{2+} of -0.76 V, Fe^{2+} of -0.44 V, Cu^{2+} of $+0.43$ V, and F_2 has a $E°$ of $+2.87$ V. When the value of the standard reduction potential is greater than another element's standard potential, the first element has a higher tendency to be reduced by accepting electrons from the latter. For instance, F_2 has a high tendency to reduce therefore being a good oxidizing agent, while Zn^{2+} would be more easily oxidized being a good reducing agent. Li^+ would be oxidized by all other elements with higher $E°$, while Cu^{2+} is reduced in combination with Zn^{2+} which oxidizes, as already seen in the previous example.

Electrochemical cells can be combined to form a *battery*, or *accumulator*, an electrochemical device capable of converting electrical energy into chemical energy during charging, storing it after the charging cycle, and converting it back to electricity during the discharging cycle. The total energy $E(t)$ stored in a battery is given by:

$$E(t) = E_{\text{in}} + \int_0^t U_B(t)\, I_B(t)\, \mathrm{d}t \tag{1.5}$$

where E_{in} is the energy stored in the battery at initial conditions, and U_B and I_B the voltage and current in the battery, respectively. Knowing the maximum energy E_{max} that the battery is capable of storing, the battery *state of charge* (SOC) is defined by:

$$\text{SOC} = \frac{E(t)}{E_{\text{max}}} \tag{1.6}$$

Conventionally, during charging the sign of the current I_B provided to the battery is positive, while it is negative during battery discharge.

The battery voltage $U_B(t)$ can be obtained by the following equation:

$$U_B(t) = (1 + \alpha t)\, U_{B,0} + R_i\, I_B(t) + K_i\, Q_R(t) \tag{1.7}$$

where t is time, α is the self-discharge (in Hz), $U_{B,0}$ is the open-circuit voltage at $t = 0$, R_i is the battery internal resistance, K_i is a coefficient taking into account battery polarization phenomena (in Ω/h), and $Q_R(t)$ is the accumulated charge (in Ah).

---

[4] The absolute hydrogen electrode potential is estimated as being 0.41 V at 25°C.

1.2 Reaction Kinetics

A chemical reaction can be generally described by the following:

$$j\,A + k\,B \rightarrow l\,C + m\,D \tag{1.8}$$

where A, B, C, and D are the reacting chemical species and j, k, l, and m are the respective stoichiometric coefficients.

From the *law of mass action* (by Guldberg and Waage) at equilibrium and fixed temperature, the constant of reaction K is given by:

$$K = \frac{a_C^l\, a_D^m}{a_A^j\, a_B^k} \tag{1.9}$$

where a_i is the *activity* of the reacting substances, calculated as the concentration in the solution or the partial pressure at equilibrium. The activity can be given, for instance, by:

$$a_i = \frac{p_{i,\text{eq}}}{p_{\text{stc}}} \tag{1.10}$$

where $p_{i,\text{eq}}$ is the partial pressure of reactants i at equilibrium and p_{stc} is the pressure in standard conditions.[5]

In case of an electrochemical reaction, the potential E of the cell in which the reaction takes place is given by the *Nernst's equation*:

$$E = E^0 - \frac{RT}{z\,F} \log Q \tag{1.11}$$

where Q is the reaction quotient, E^0 is the potential in standard conditions, R is the universal constant of gases, z is the number of electrons involved in the electrochemical reaction, and F is the Faraday's constant.

According to *Le Chatelier's principle*, every chemical system reacts to an externally imposed modification to minimize the effect of the change. Therefore, if there is a perturbation, the system will shift either toward the products or to the reactants to counteract the change. At equilibrium, Q equals K.

1.3 Reaction Thermodynamics

The enthalpy variation ΔH in a chemical reaction is defined as the difference between the sum of the enthalpies of formation of the products and that of the enthalpy of formation of the reactants:

$$\Delta H = \sum H_{f,\text{products}} - \sum H_{f,\text{reactants}} \tag{1.12}$$

A chemical reaction is *exothermic* when thermal energy is released to the environment; in this case, the enthalpy difference is negative. When the difference is positive,

---

[5] Standard conditions refer to a pressure of 0.1 MPa and a temperature at 25°C. For a chemical element, the standard state is the condition it assumes at standard pressure and temperature.

the reaction is *endothermic* and occurs only when it absorbs energy from the external environment.

The energy effectively available to generate work is what remains from ΔH after removing the product of temperature T and of entropy S. The resulting state function is the *Gibbs' free energy*. This function is important because in nature, these transformations usually occur at a constant pressure and temperature rather than at a fixed volume. It is defined as:

$$G = H - TS \tag{1.13}$$

The Gibbs' free energy determines if a chemical reaction will happen spontaneously at a given temperature, since a spontaneous reaction occurs only when the variation of free energy ΔG is negative. The infinitesimal difference of Gibbs' free energy can be expressed as:

$$dG = dH - TdS - SdT \tag{1.14}$$

When T is constant:

$$dG = dH - TdS \tag{1.15}$$

with :

$$dH = dU + pdV + Vdp \tag{1.16}$$

where U is the internal energy, p the pressure, and V the volume.

If p is also constant:

$$dG = dU + pdV - TdS \tag{1.17}$$

If a thermodynamic transformation happens between two infinitely near equilibrium states, the first principle of thermodynamics can be described as:[6]

$$dU = \delta Q - \delta L \tag{1.18}$$

From which it is derived that:

$$dG = \delta Q - \delta L + pdV - TdS \tag{1.19}$$

In an reversible transformation, δQ equals TdS; therefore, the previous formula can be reduced to:

$$dG = - \left(\delta L - pdV \right) \tag{1.20}$$

For this reason, in electrochemical cells, all the work that is not lost as volume change is available as electric work. In a reversible transformation, the work calculated as variation of Gibbs' free energy is ideal work.

---

[6] While the internal energy is an exact differential because it depends only on the initial and final states, Q and W are not state functions, therefore their integral depends on the cycle. For this reason, the symbol δ is used instead of d to indicate that it is not the exact differential but rather an infinitesimal quantity.

When a spontaneous redox reaction in an electrochemical cell occurs, dG must be negative, as given by:

$$dG = nF\,E°ominus \tag{1.21}$$

where n is number of moles of electrons and F is the Faraday constant. For a spontaneous reaction to occur, $E°$ given by $E°_{\text{cathode}} - E°_{\text{anode}}$ must be positive.

1.4 Efficiency

Usually, three types of efficiency are provided for electrochemical batteries.

The *coulombic efficiency* (also referred to as *charge acceptance, charge efficiency, Faraday efficiency,* and *faradaic efficiency*) is the ratio between the charge (in Ah or Coulomb) that has been charged in the battery, and the charge that is made available for discharge:

$$\eta_{\text{coul}} = \frac{Q_{\text{out}}}{Q_{\text{in}}} \tag{1.22}$$

If all electrons are available for useful work in the load, coulombic efficiency can be 100%. If electrons participate in side reactions or recombine before being made available for useful work, coulombic losses occur and manifest as heat or some chemical byproducts in the cell. Self-discharge is an example of coulombic losses.

The *energy efficiency* is the ratio between the energy (in kWh) that has been charged in the battery, and the energy that has been available during discharge:

$$\eta_{\text{en}} = \frac{E_{\text{out}}}{E_{\text{in}}} \tag{1.23}$$

Energy efficiency is not 100% due to the thermodynamic irreversible energy conversion happening in real-world processes. Energy is converted part as useful work, part as heat.

Another index related to useful energy-conversion losses is the *voltaic efficiency*, namely, the ratio between the average voltage (in V) during charge, and the average voltage during discharge:

$$\eta_{\text{vol}} = \frac{U_{\text{ave,out}}}{U_{\text{ave,in}}} \tag{1.24}$$

An example of lower than 100% voltaic efficiency is the overpotential, the phenomenon occurring when charging voltage is higher than discharging voltage, or the difference between the theoretical voltage and the real cell voltages resulting during actual cell operations.

1.5 Battery Specifications and Operating Parameters

The main specifications for electrochemical batteries are:

- *Nominal capacity* C_n (or *rated capacity*) is the value of the charge (in Ah or C) available in the battery at a given discharge rate or rated discharge time τ

- *Nominal energy* E_n (or *rated energy*) is the value of the energy available in the battery (in Wh) at a given discharge rate or rated discharge time τ

- *C-rate* is the rate of charge or discharge of a battery normalized on its nominal capacity. A battery discharging at a C-rate of τ will deliver its nominal rated capacity in τ h. For instance, if the rated capacity is 2 Ah, a discharge rate of 1C (or C/1) corresponds to a discharge current of 2 A, which is the current that will be delivered in a 1 h period; a rate of 2C is a current of 4 A, that will discharge the battery in 30 min; a rate of C/2 is a current of 1 A, capable of discharging the battery in 2 h

- *E-rate* is the rate of charge or discharge of a battery over its nominal energy. A 1E rate means that in 1 h, the full battery energy will be discharged

- *Rate-dependent capacity* is the different capacity (in Ah or C) that some battery technologies provide as a function of the charge and discharge currents. Depending on current value for instance and due to internal battery chemical kinetics, not all available capacity can be discharged to the load or be charged inside the battery

- *Specific energy or power* is the *gravimetric density* of energy or power in Wh/kg or W/kg

- *Energy or power density* is the *volumetric density* of energy or power in Wh/l or W/l

- *Self-discharge rate* is the rate at which a battery loses its rated capacity when standing idle under open circuit conditions due to parasitical chemical reactions or current leakages occurring inside the battery

- *Internal resistance* is the internal resistance of the battery during operations

- *Maximum state of charge* (SOC_{max}) is the maximum level of charge (in %) to which the battery can be safely charged

- *Minimum state of charge* (SOC_{min}) is the minimum *state of charge* (SOC) level of charge (in %) to which the battery can be safely discharged

- *Cycle life* (or *cycle number*) is the guaranteed or total number of charge and discharge cycles from SOC_{max} to SOC_{min} that a battery can sustain before losing a predetermined performance, whether on capacity and/or efficiency

The main operating battery parameters are:

- State of Charge (SOC) is the capacity available in the battery (in percentage between 0 and 100%). SOC measurement is essential for a correct evaluation of the remaining battery capacity and correct functioning

- *State of health* (SOH) is the status of the condition of the battery (in percentage between 0 and 100%) compared to its initial nominal conditions. Typically, a battery SOH will be 100% at the time of manufacture and will decrease over time, type of use, and the number of cycles at which the battery will be subject to

- *Depth of discharge* (DOD) is the percentage of battery capacity that has been discharged

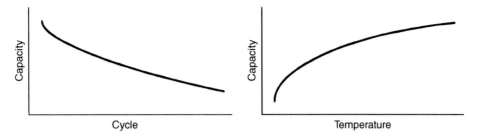

FIGURE 1.2   Capacity as a function of cycle numbers and temperature.

- *Open circuit voltage* (OCV) is the voltage measured at the battery terminals when disconnected from the load. Normally, OCV depends from SOC, being the highest when SOC is equal to 100%. One possible technique to compute SOC at the battery terminals is to correlate OCV with SOC
- *Terminal voltage* is the voltage at the battery terminal when load is applied

1.6   Behavior of Electrochemical Batteries and the Peukert's Law

Electrochemical batteries, even if belonging to different technologies, tend to share a common behavior along their lifetime or during charge and discharge cycles within their daily operations [2, 3]. The graphs in Figs. 1.2 and 1.3 show examples of typical capacity and voltage variations as a function of time, temperature, current, and SOC. These curves do not apply to all types of battery technologies, but are representative of common behaviors of a large number of electrochemical storage devices, and provide very useful insights on the characteristics of the underlying functioning of batteries.

The usable capacity of electrochemical batteries decreases when the battery has aged after a number of charge and discharge cycles (Fig. 1.2, left); the higher the number of cycles, the lower the available maximum charge of the battery. Several can be the causes for such decrease, from decay in the internal chemistry, to mechanical changes in electrodes, creep of the cell components, and other phenomena.

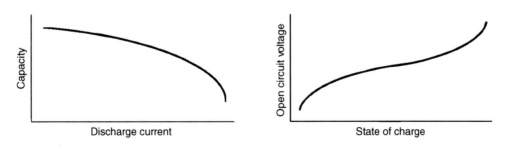

FIGURE 1.3   Capacity as a function of discharge current and open circuit voltage as a function of state of charge.

Usable capacity is also a function of temperature, with lower charge available in the battery under cooler environmental conditions. Normally, cold temperatures tend to affect chemical kinetics therefore reducing available charge in the electrochemical cell (Fig. 1.2, right).

A very important dependency of a battery usable capacity is with the amount of electrical current that is drained from the battery during discharge (with a similar behavior during charge): if the discharge current increases, many technologies show a decrease in the usable capacity (Fig. 1.3, left). The *Peukert's law* provides an empirical relation to assess the capacity of a battery as a function of discharge currents. The *Peukert's capacity* C_P, representing the real capacity that can be obtained from the battery at the discharge current i_{dis}, is given by:

$$C_P = C_n \left(\frac{C_n}{i_{dis} t_n} \right)^{k-1} \tag{1.25}$$

where k is the *Peukert's coefficient*, computed from experimental data for different battery technologies. If k is equal to one, regardless of whether the discharge current varies, the capacity made available by the battery is the same. Normally, $k > 1$ and the capacity available decreases when discharge currents increase. Many electrochemical batteries exhibit a behavior that can be described by Peukert's coefficients k higher than one. Nominal capacity C_n can therefore be only an indicative parameter of battery capability. Normally, all manufacturer's data-sheets provide curves of capacity as a function of charge and discharge currents, that are needed by system designers to correctly dimension the overall energy storage system (ESS).

Open-circuit voltage is a function of the battery SOC (Fig. 1.3, right); the nonlinearity especially in the low and high tails of the SOC range is due to the polarization effects when the battery is close to full charge or full discharge and depend on the electrochemistry of each battery technology.

Another typical behavior for electrochemical batteries is the phenomenon linked to coulombic losses is the self-discharge that reduces usable capacity over the period of time during which the ESS is kept idle and uncharged.

Finally, the dynamic characteristics of batteries during the transients of charge and discharge can have very important design implications, especially in operating modes when charge and discharge cycles can occur within very short periods of time, as in the case of nonstationary applications for electric vehicles. From the curves that log the variation of output voltage to changes in charge and discharge currents, the main time constants of the charge and discharge processes can be computed and used in modeling the ESS during transients or longer periods of time.

1.7 Charge and Discharge

Depending on battery technology and manufacturer's specifications, charge and discharge techniques can vary significantly from one battery to the other. Typically though, batteries can be charged or discharged using a combination of constant current, constant voltage, constant power modes. Depending on the type of battery and their adoption in the industry, a set of norms and regulations have been approved by regulatory bodies

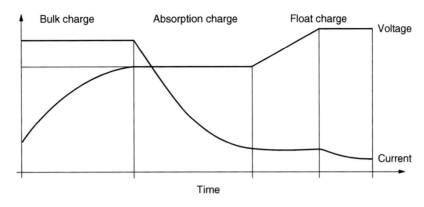

FIGURE 1.4 Constant current, constant voltage charge mode example.

and must be complied with by battery charger manufacturers (i.e., DIN 41773, DIN 41774 for lead-acid batteries).

Charging is the process of replenishing the battery cell capacity by storing energy and increasing capacity from an initial value to a final value higher than the previous. For instance, if a cell has been used and its capacity drained to SOC_{min}, a full charge cycle increases the SOC to cell SOC_{max}.

Normally, the charging process entails three steps.[7]

The *bulk charge* is the initial step during which the largest portion of the cell capacity is replenished by the charger unit. During bulk charge, the constant current mode or constant power mode are normally employed. In this phase, the cell heats up due to internal resistance phenomena. Charging must therefore be performed without overheating the cell, where the level of charge applied to the cell without overheating is called the cell *natural absorption rate*.

The *absorption charge* (also called *saturation charge*) is the phase that typically starts when SOC has reached a value in the range of 80%. From now on, a constant voltage is applied to the terminals of the cell, while current gradually decreases due to the fact that the cell is absorbing less and less capacity, until the cell reaches its full capacity. No overheating occurs during this phase since current is not high enough to dissipate as heat on the internal resistance. This phase takes normally longer than bulk charge. In this phase, the voltage applied to the cell can be higher than its nominal value, depending on cell technology.

The *float charge* (also known as *trickle charge*) is the final step that normally begins when SOC is equal to 95%, and brings it to 100%, while this stage can be entered into directly at SOC_{max}. Float charge is used to maintain full SOC, although it can be avoided and the cell left in float, or idle, mode for an unspecified period of time. During float charge, the voltage can be maintained constant while current shows a trickle behavior, dropping small quantities of capacity into the cell.

An example of a constant current and constant voltage charge curve is provided in Fig. 1.4. The bulk charge phase is performed at constant current, and the absorption

---

[7] Another cell-charging process is "cell equalization," namely, the process of making each cell SOC nearly the same in batteries comprising of multiple cells. This process will be discussed in Chap. 6.

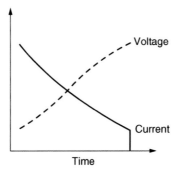

FIGURE 1.5
Constant power
charge mode
example.

charge phase at constant voltage. The float charge phase is operated as a two step, one at constant current, the second at constant voltage, until charger logic interrupts the process when SOC_{max} is detected.[8]

In *constant power charge mode*, current and voltage vary as shown in Fig. 1.5.

Increasing the charge current normally does not reduce charging time to reach SOC_{max}. If the bulk charge is shortened, absorption charge will take longer. In cases when reaching a percentage of total charge is sufficient, fast charging can anyway be considered as an option. This can be useful, for instance, in nonstationary applications where charge time must be kept down to the minimum. As an additional benefit, avoiding full charge can increase life of certain battery technologies.

Discharging is the process of using the battery cell capacity by using energy and decreasing capacity from an initial value to a final value lower than the previous. For instance, if a cell has been fully charged and its capacity has reached SOC_{max}, a full discharge cycle decreases the SOC down to cell SOC_{mon}.

Typically, discharge occurs at constant current at varying C-rates, or in constant power mode. When a constant current is drained from the ESS, the bus voltage tends to decrease as a sign of reducing capacity. At high C-rates, the full available capacity cannot be used. Typical discharge curves, as a function of different C-rates, are shown in Fig. 1.6.

A set of specifications is therefore given to comply with during design:

- *Nominal voltage* (U_n), is the voltage of a fully charged battery when delivering rated capacity at a specific discharge rate
- *Cut-off voltage* is the voltage at the terminals of the battery at value at SOC_{min}, when the battery discharge is stopped to avoid battery damages or because there is no more capacity left in the battery
- *Float voltage* is the voltage kept at the terminals of the battery at SOC_{max} to compensate for self-discharging
- *Recommended charge current* is the current with which the battery is charged in constant current mode before switching to constant voltage charging mode during the last 20–30% of SOC in order to reach SOC_{max}

---

[8] This operational strategy for cell charging can be found in technical documents as the "IUIUa" curve.

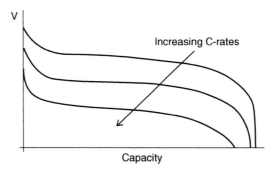

FIGURE 1.6 Voltage as a function of capacity at different discharge currents.

- *Recommended charge voltage* is the voltage with which the battery is charged in constant voltage mode to reach SOC_{max}
- *Maximum continuous charge/discharge current* is the maximum current with which the battery can be charged or discharged continuously without damaging the battery itself
- *Maximum pulsed charge/discharge current* is the maximum allowed current during pulsed duty cycles (usually set at 30 s) that do not damage the battery

1.8   Solid-State Batteries

Solid-state batteries are electrochemical storage technologies where, the internal chemistry is in solid state at ambient temperature and pressure conditions.

Within the rechargeable electrochemical batteries that will be covered in this section are:

- lead-acid
- nickel based
- lithium based
- zinc-air
- molten salts
- aqueous Ions

Due to the many differences that can be found in the market, resorting to manufacturer's data-sheets is advised for a more complete analysis of each technology characteristic curves and operating parameters.

1.8.1   Modeling of Solid-State Batteries

Models are useful to design and evaluate battery performances when used in conjunction with renewable energy (REN) sources, load profiles, other storage systems, and strategies for control of all the possible applications that ESS can be used for. There are many possible ways to conceive models to describe and simulate the behavior of electrochemical batteries.

Electrochemical models correlate the physics and chemistry of the battery with its electrical functioning, although models are often consisting of sets of partial differential equations that are difficult to interpret, need long computation times to solve, and are difficult to generalize for other battery technologies.

Mathematical models use statistical tools or empirical equations, often based on regressions over experimental data, to obtain precise results but present the same difficulties of electrochemical models.

Electrical models, though less precise than the other two modeling strategies, have some benefits that can make their use preferable in many practical applications: they can be more intuitive and can give a better understanding of how the battery works, are easier to generalize, provide a good description of steady and transient states, and are fast in providing results when used in simulation software.

Kinetic Battery Model

According to Eq. (1.25), if a battery is discharged at rates higher than nominal, the capacity that is made available to the load is not the total nominal capacity that can be extracted at nominal rates. This occurs in some electrochemical technologies more than in others, and is due to the internal chemistry dynamics peculiar to each technology. For some batteries then, the only way to extract the highest efficiency is to design them for applications where long charge and discharge times are acceptable.

A model that has been devise to describe and simulate this behavior is the *Kinetic Battery Model* (KBM) [4, 5], where a battery energy content is, by analogy, assimilated to the content of two reservoirs connected to each other with one duct, and only one of the two reservoirs has an additional pipe connecting it to the outside load (Fig. 1.7).

The total energy content in the system is equivalent to $C_n = y_1 + y_2$, while each reservoir energy content is given by $y_2 = (1 - c) \times h_2$ for the "bound charge" reservoir, and

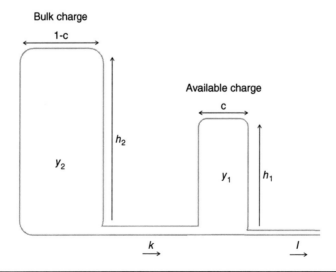

FIGURE 1.7 Schematic of the Kinetic Battery Model (modified from [5]).

by $y_1 = c \times h_1$ for the "available charge" reservoir. The flow of charge from one reservoir to the other depends on the parameter k and on the relative heights h_1 and h_2; the charge rate flowing from the "available charge" reservoir to the load depends on the parameter I.

Not all total charge is capable of flowing to the load, but a portion of it will have to flow previously from the "bound charge" reservoir to the "available charge" reservoir. If the flow rate I is too high, the "available charge" reservoir will be depleted before the "bound charge" reservoir can replenish it. At slow discharge rates, all "bound charge" will be converted to "available charge," with an increase in the available capacity that will be equal to nominal. The same will occur during charge. Relative amount of "bound" and "available" energy will depend on the relative rates of charge and discharge cycles, if these cycles have not been sufficiently slow to completely refill the reservoirs.

The charge variation over time t in the two reservoirs is given by the following set of differential equations [5]:

$$\begin{cases} \dot{y}_1 = -i(t) + k(h_2 - h_1) \\ \dot{y}_2 = -k(h_2 - h_1) \end{cases} \tag{1.26}$$

with the initial conditions $y_1(0) = c \cdot C_n$ and $y_2(0) = (1-c) \cdot C_n$. By applying a coordinate transformation using $\delta = h_2 - h_1$, and $\gamma = y_1 + y_2$:

$$\begin{cases} \dot{\delta} = \dfrac{i(t)}{c} - k'\delta \\ \dot{\gamma} = -i(t) \end{cases} \tag{1.27}$$

with the boundary (initial) conditions: $\delta(0) = 0$ and $\gamma(0) = C_n$. The differential equations are independent and, under the hypothesis of constant discharge current, yield:

$$\begin{cases} \delta(t) = \dfrac{I}{c} \dfrac{1 - e^{-k't}}{k'} \\ \gamma(t) = C_n - It \end{cases} \tag{1.28}$$

When a battery discharges, part of the charge in the "available charge" reservoir flows to the external load, and part of the charge of the "bound charge" flows to replenish the "available charge" reservoir. Depending on discharge rate, it can happen that the "available charge" reservoir empties and discharging is stopped and load disconnected. After a recovery period, sufficient charge from the "bound charge" will have replenished part of the "available charge" reservoir, making it possible to start a new discharge process without necessarily recharge the battery from a external power source. The recovery after an idle time t_i from a discharge current I that lasted for a period t_{dis} can be computed from the model as:

$$\delta(t_i) = \frac{I}{c} \frac{e^{-k't_i(1-e^{-k't_{dis}})}}{k'} \tag{1.29}$$

The trend followed by voltage an the effect described by the KBM during charge of discharge is due to irreversible processes occurring in the chemistry of the battery. These irreversibilities manifest as losses that can be singled-out as:

- *activation losses* caused by activation energy that must be spent for the reaction to occur

- *ohmic losses* due to electrical resistance to the flow of electrons caused by the materials composing the various battery layers and interfaces, resistance that is proportional to the current density

- *concentration losses* (or *mass transport losses*) arising from the nonhomogeneity of the concentration of reactants inside the battery and the noninstantaneous transport of ions on the surface of the electrodes, occurring especially at high discharge rates

Shepherd's and Unnewehr's Model Equation

The *Shepherd's equation* is used as a modeling tool that describes the electrochemical behavior of the battery directly in terms of voltage and current [6]. It can be used together with Peukert's equation to compute battery voltage and SOC as a function of the power draw [7]:

$$U(t) = U_0 - R_i\, i(t) - K_i\, Q_0 \frac{1}{1 - \int_0^t i(t)\mathrm{d}t} \tag{1.30}$$

where $U(t)$ is the battery voltage at time t, U_0 is the open circuit voltage at full charge, R_i is the internal resistance, $i(t)$ is the current as a function of time, K_i is the polarization resistance, and Q_0 is the nominal battery capacity.

In order to estimate the unknown parameters R_i and K_i that depend on battery technology, a number N of measurements are taken and a curve fitting procedure is performed. The set of equations that are obtained are the following:

$$U_k = U_0 - R_i\, i_k - \frac{\mu}{\mathrm{SOC}_k}, \forall k = 1, \ldots, N \tag{1.31}$$

where U_k are the N measures on battery terminal voltage $U(t)$, SOC_k are the measures on SOC, i_k are the measures on current, and μ is the curve fitting parameter. By employing a matrix notations:

$$\begin{bmatrix} U_1 \\ U_2 \\ \vdots \\ U_N \end{bmatrix} = \begin{bmatrix} 1 & I_1 & \ldots & 1/\mathrm{SOC}_1 \\ 1 & i_2 & \ldots & 1/\mathrm{SOC}_2 \\ \vdots & \vdots & \ddots & \vdots \\ 1 & i_N & \ldots & 1/\mathrm{SOC}_N \end{bmatrix} \begin{bmatrix} U_0 \\ R_i \\ \mu \end{bmatrix}$$

By defining the matrix H:

$$H = \begin{bmatrix} 1 & I_1 & \ldots & 1/\mathrm{SOC}_1 \\ 1 & i_2 & \ldots & 1/\mathrm{SOC}_2 \\ \vdots & \vdots & \ddots & \vdots \\ 1 & i_N & \ldots & 1/\mathrm{SOC}_N \end{bmatrix}$$

with the unknown vector $\bar{x}$ given by:

$$\bar{x} = \begin{bmatrix} U_0 \\ R_i \\ \mu \end{bmatrix}$$

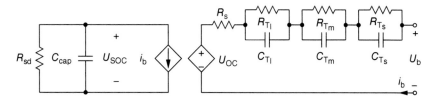

FIGURE 1.8 Electrochemical battery equivalent electrical model.

obtained by solving for least-square method as:

$$\vec{x} = (H^T H)^{-1} H^T Y \tag{1.32}$$

In hybrid electric vehicles (HEV), the charge–discharge patterns for batteries never occur close to the SOC limits, thus the *Unnewehr's equation* holds by accounting for such simplification as per:

$$U(t) = U_0 - R_i \cdot i_k(t) - \mu \cdot \text{SOC}(t) \tag{1.33}$$

and solved as the Shepherd's equation.

Equivalent Electrical Model

A practical way to model the behavior of electrochemical batteries is by employing equivalent electrical models based on circuit theory. One of the more thorough electrical models capable of simulating and assessing the battery run-time and its $U - I$ performance with sufficient precision and meaningfulness is shown in Fig. 1.8. The circuit is divided into two subcircuits: the first (on the left-hand side of Fig. 1.8) describing the battery lifetime behavior, the second (on the right-hand side of Fig. 1.8) describing the run-time variations of $I - U$ at transient and static conditions. The two circuits are cross-correlated by the voltage generator of the run-time circuit $U_{OC} = f(U_{SOC})$, function of the output of the battery lifetime circuit.

In the battery-lifetime circuit, the C_{cap} is the battery capacity that varies over time, the R_{sd} is a resistance that models the phenomenon of self-discharging of the capacity when the battery is kept in idle mode and disconnected from the charging source. The current generator i_b is the current that flows in and out of the battery as a function of U_b. In the run-time RC circuit, the voltage generator models the voltage at the terminals of the battery as a function of the SOC, R_s accounts for the voltage drop in steady-state conditions, while the three RC parallel circuits model the transient response of the battery with time constants, long, medium, and short after a transient in the charge–discharge current has occurred.

All the parameters in the model are relatively easy to be extracted from experimental data provided by graphs obtained from measurements and data-logs from experimental setups.

All the parameters in the model are multidimensional functions of SOC, current, voltage, temperature, cycle number, and aging. It is still possible to achieve good estimates by reducing the dependency of some parameters by linearization or by ignoring the effect of variables the impact of which is negligible on the other parameters (i.e., self-discharge during daily charge–discharge operations).

For example, in the model the usable capacity C_{cap} can be given by:

$$C_{cap} = 3600 \times \text{Capacity} \times f_1(\text{Cycle}) \times f_2(\text{Temp}) \tag{1.34}$$

where C_{cap} is the nominal capacity, f_1 (Cycle) and f_2 (Temp) are correction coefficients depending on cycle number and temperature.

The internal impedance of the battery is approximated by the four resistances R_S, R_{Tl}, R_{Tm}, and R_{Ts}, and the three capacitors C_{Tl}, C_{Tm}, and C_{Ts}. Such impedance and the current i_b, set by the current controlled generator, give the variation of C_{cap} during battery charge or discharge operations with SOC (from U_{SOC}) varying dynamically. All the time constants can be estimated by evaluating the transient response in the $U - I$ curves from laboratory data.

1.8.2   Lead Acid

Lead-acid battery technology dates back to many decades and is known for its reliability and affordability. Its use is mostly for UPS and power quality, with a 70% use in vehicles and telecom applications. Due to its inherent disadvantages, mostly related to safety issues from emission of inflammable gases and acids during charge and discharge operations, the original design has been improved over the years into *valve regulated lead-acid* (VRLA) batteries, packaged to avoid emissions and requiring very few maintenance.

The *absorbent glass mat* battery is a type of VRLA where the electrolyte is absorbed in a fiberglass matrix allowing for higher purity levels since the electrolyte does not need to be capable of sustaining its own weight inside the battery vessel. Compared with traditional batteries, this device benefits from a lower internal resistance, endures higher operation temperatures, operates with lower discharge rates and possesses higher power densities. This battery is therefore suitable to be used also in electric vehicles.

The *gel battery* is a VRLA where the electrolyte is in a gel form with the sulfuric acid electrolyte mixed with silica fume. This nonliquid feature allows the battery to avoid problems of leakage or evaporation. A gel battery has no need to be kept upright and is also very resistant to external impacts, shocks, vibrations, and high operation temperatures.

In VRLA batteries, the charge–discharge processes are generated by the movement of lead ions from one electrode to the other. The anode is made of lead while the cathode is made from porous lead-oxide.

During the discharge cycle, the anode lead is attacked by the sulfuric acid resulting in the formation of lead ions (Pb^{2+}) and the correspondent electrons (e^-) that flow in the external electric circuit with this oxidation reaction:

$$Pb \rightarrow Pb^{2+} + 2\,e^- \tag{1.35}$$

At the anode, lead sulfate and hydrogen ions are formed following this reaction:

$$Pb^{2+} + HSO_4^- \rightarrow PbSO_4 + H^+ + 2\,e^- \tag{1.36}$$

At the lead-oxide electrode, the electrons e^- from the external electric circuit react with Pb^{4+} with the reduction reaction:

$$Pb^{4+} + 2\,e^- \rightarrow Pb^{2+} \tag{1.37}$$

At the cathode, lead sulfate and water are produced as per:

$$PbO_2 + HSO_4^- + 3\,H^+ + 2\,e^- \rightarrow PbSO_4 + 2\,H_2O \qquad (1.38)$$

The charging cycle is similar to discharging except that it occurs in the reversed direction and restores the initial electrolyte charge.

The output voltage is around 2 V as per the sum of the absolute values of the reaction potentials at the anode and at the cathode (-0.36 and 1.69 V in standard conditions, respectively). The batteries therefore need to be connected in stacks by connecting several single battery units in series and the total number of the unit depends on the desired output voltage. Normally, for nonstationary uses, the output voltage is 12 V, while for nonstationary applications the voltage can vary from 24 to 220 V.

The typical model for Lead Acid batteries is the KBM model, but it is also possible to use Shepherd's like equations to model the behavior of the battery during charge and discharge cycles. Discharging can indeed be modeled by the following equation:

$$U_{\text{dis}} = E_0 - R\,i - K_c \frac{Q}{Q - \int i\,dt} \left(\int i\,dt + i^* \right) + E(t) \qquad (1.39)$$

where:

- U_{dis} is the battery voltage
- E_0 is the nominal battery voltage, usually taken as the voltage measured at the end of the linear zone of the discharge curve
- R is the internal resistance
- i is the discharge current
- K_c is a polarization constant ($VA^{-1}h^{-1}$), meant to improve modeling of the nonlinearity of the relationship of open circuit voltage with SOC
- Q is the battery nominal capacity
- $\int i\,dt$ is the actual capacity
- i^* is a current used to model low-frequency battery dynamics

In lead-acid batteries, the exponential region of the discharge curve has an hysteresis behavior that makes the discharge exponential region different from the charge exponential region, independently from SOC. $E(t)$ is a function describing the behavior of the battery in the exponential region of the discharge/charge curve, and can be modeled during discharge as:

$$E(t) = B\,|i(t)|\,E(t_0) \qquad (1.40)$$

while during charge:

$$E(t) = B\,|i(t)|\,\left(E(t_0) + A \right) \qquad (1.41)$$

where B is the inverse of the time constant of the curve in the exponential region ($A^{-1}h^{-1}$), $E(t_0)$ the initial value (V), and A is the exponential region amplitude.

While nearing the end of charge (EOC), the voltage increases very fast; this behavior is modeled using a polarization resistance K_r, which increases when battery reaches EOC.

Specific energy (Wh/kg)	20–50
DOD (%)	40–70
RT efficiency (%)	70–85
Typical cycle number	700–1000
Self-discharge (% monthly)	2
Response time	10 s

TABLE 1.1   Lead-Acid Battery Main Data

Charge can therefore be described by the following equation:

$$U_{ch} = E_0 - Ri - K_c \frac{Q}{\int i \, dt - 0.1 \times Q} i^* - K_r \frac{Q}{Q - \int i \, dt} \int i \, dt + E(t) \qquad (1.42)$$

All the parameters used in these equation are extracted from the discharge and charge curves from laboratory data.

Energy density is in the range of 20–50 Wh/kg with DOD in the range of 50%, and efficiency between 75 and 85%. Unfortunately, such batteries tend to have very low cycle numbers and therefore need to be replaced earlier than other battery technologies. DOD is kept low to avoid damages to the electrodes and rapid degradation.

The main data for lead-acid batteries are provided in Table 1.1.

1.8.3 Nickel Based

Nickel, symbol Ni and atomic number 28, is an element belonging to the transition metal class with a crystalline structure, and oxidation states +2 and +3. This element is used in association with cadmium, metal hydrides, or hydrogen, in the following nickel-based batteries:

- nickel cadmium
- nickel-metal hydride
- nickel-hydrogen

Nickel-Cadmium

Cadmium, symbol Cd and atomic number 48, is an element belonging to the transition metal class, and oxidation state +2 (rare +1). In the *nickel-cadmium battery* (Ni-Cd), Cd is used in combination with Ni where the cathode is a nickel oxyhydroxide electrode, the anode is a Cd electrode, and the electrolyte is KOH. Small-sized Ni-Cd batteries are constructed in a spiral way (the so called *jelly-roll design*), while larger battery assemblies employs the usual stack configuration.

The reactions taking place in the cell are the following:

$$2\,NiO(OH) + Cd + 2\,H_2O \rightarrow 2\,Ni(OH)_2 + Cd(OH)_2 \qquad (1.43)$$

where discharge occurs in the left-to-right direction, and charge in the reverse direction.

Large capacity Ni-Cd batteries are usually built with a vent valve in order to avoid over-pressure inside the cell casing; at high charge and discharge rates in fact, hydrogen and oxygen that develop inside the cell do not have the time to recombine and, if not vented, would reduce efficiency and cause safety issues.

Specific energy (Wh/kg)	35–60
DOD (%)	90–100
RT efficiency (%)	60–90
Typical cycle number	2000–3000
Self-discharge (% monthly)	5–10
Response time	s

TABLE 1.2   Ni-Cd Battery Main Data

Advantages of Ni-Cd batteries are the very high discharge rates and the possibility to be used in environments where thermal and mechanical stresses are high. They are used, for instance, in battery-powered tools and in the aerospace industry.

Important disadvantages are that Ni-Cd are toxic and harmful for the environment. *Memory effect*, namely, the loss of nominal capacity, can occur if charging starts before the battery has fully discharged.

Specific energy typically ranges within 35–60 Wh/kg, with DOD 90% and cycle life around 3000. Self-discharge can be for some applications not acceptable since it can range at a fairly high 10% per month. Response times are in the range of seconds.

As for lead-acid batteries, the Shepherd's equations can be applied to Ni-Cd. During discharge:

$$U_{\text{dis}} = E_0 - Ri - K_c \frac{Q}{Q - \int i\,dt} \left(\int i\,dt + i^* \right) + E(t) \tag{1.44}$$

where the parameters are described in Sec. 1.8.2, and $E(t)$ is given in Eq. (1.40).

After the battery has reached the EOC, the voltage decreases slowly as a function of the charge current. This occurrence is modeled by using the absolute value of the actual capacity of the battery. Therefore, during charge:

$$U_{\text{ch}} = E_0 - Ri - K_c \frac{Q}{|\int i\,dt| - 0.1 \times Q} i^* - K_r \frac{Q}{Q - \int i\,dt} \int i\,dt + E(t) \tag{1.45}$$

Main battery data are provided in Table 1.2.

Nickel-Hydrogen

The *nickel-hydrogen battery* (Ni-H$_2$) employs diatomic hydrogen stored by compression in its gaseous form instead than the cadmium anode, making it a hybrid battery between a Ni-Cd and a fuel cell (see Sec. 3.2.3). The battery employs a sintered porous nickel hydroxide (NiOH$_2$) cathode, alkaline potassium hydroxide (KOH) for the electrolyte, and an anode that is coated with catalyst as in fuel cell technology (i.e., platinum). The separator can be zirconia or asbestos, although the latter has mostly been used in the first production batches.

In discharge mode, hydrogen is oxidized into water and the nickel oxyhydroxide cathode is reduced to nickel hydroxide. Water concentration, and therefore KOH concentration, remains the same since water is consumed and produced at the different electrodes in the same quantity.

The greatest advantage of this technology is the very long cycle life which is over 15,000 cycles at a DOD of 80% and efficiency around 85%. The main use of this battery has been mostly in the aerospace industry. The cell can sustain overcharging and reverse

Specific energy (Wh/kg)	50–75
DOD (%)	90–100
RT efficiency (%)	85
Typical cycle number	15,000–20,000
Self-discharge (% monthly)	> 50%
Response time	< 1 s

TABLE 1.3   Ni-H Battery Main Data

polarity operation without damages. The SOC can be inferred from the pressure of hydrogen in the battery, which is in fact a pressurized container.

One disadvantage is the very high self-discharge rate, but intensive use makes the issue not significant.

Table 1.3 summarizes the main Ni-H battery data.

Nickel-Metal Hydride

Nickel-metal hydride batteries (Ni-MH) are a hybrid between Ni-Cd and Ni-H$_2$ batteries, where the cathode is made of NiO(OH) and the anode is a hydrogen-adsorbing alloy.

The reactions that takes place in the cell are similar to the Ni-Cd reactions but the combination of the alloy with hydrogen, resulting in the metal hydride compound, is used instead of Cd. The alloys that constitute the negative electrode of the cell are intermetallic compounds, namely combinations of rare earth elements or titanium, vanadium, and metals like cobalt, aluminum or manganese. Electrolyte is an alkaline solution, and the separator is a nonwoven polyolefin.

The charge reaction at the cathode of the Ni-MH is the following:

$$M + e^- + H_2O \rightarrow MH + (OH)^- \tag{1.46}$$

with discharging occurring in the opposite direction. The charge reaction at the anode is the following:

$$NiO(OH)_2 + OH^- \rightarrow NiO(OH) + H_2O + e^- \tag{1.47}$$

with discharging occurring in the opposite direction.

The materials composing the Ni-MH battery have the advantage of being environmentally benign, and fast charge rates, high efficiency together with good operations over a wide range of temperatures, make the use of these batteries widespread for portable electronic devices, telecommunications, and electric vehicles. Cycle life is although limited, and such batteries do not last more than 4–5 years, and memory effect can take place as in Ni-Cd batteries, although with less impact on nominal capacity.

The evaluation of the SOC is based on *coulometry*, an electrochemical technique capable of measuring the capacity flowing in charge and discharge currents while comparing them with self-discharge during idle state.

Shepherd's equations can be used for charge and discharge modeling of Ni-MH; the equations are the same as the ones described for Ni-Cd batteries [Eqs. (1.44) and (1.45)].

Table 1.4 summarizes the main Ni-H battery data.

Specific energy (Wh/kg)	50–100
DOD (%)	90–100
RT efficiency (%)	80–90
Typical cycle number	1000–2000
Self-discharge (% monthly)	5–10
Response time	s

TABLE 1.4   Ni-MH Battery Main Data

1.8.4   Lithium-Ion

Lithium, symbol Li and atomic number 3, is an element belonging to the alkali metal class, a very reactive element that, as a consequence, is found in nature always in compounds with other elements. *Lithium-ion batteries* (LIB, Li-ion) are commonly used in many different industrial, military, and consumer applications. They benefit from high energy density, high reliability, good efficiency, and large temperature operating ranges, but come at higher prices than other battery technologies like lead-acid and nickel-based.

In conventional LIB, the electrolyte is a nonaqueous solution of lithium salts (like $LiPF_4$, $LiPF_6$, $LiAsF_6$, and $LiClO_4$) in an organic solvent. The anode is usually in graphite, while the cathode can be made of metal oxides (like lithium cobalt oxide, $LiCoO_2$), crystal oxides spinels (like lithium manganese oxide $LiMg_2O_4$), or polyanions[9] (like lithium iron phosphate $LiFePO_4$). A thin sheet of special plastic separates the positive half-cell to the negative half-cell; the micro holes in the plastic let the lithium ions free to move from one half cell to the other.

During charging, the electric potential applied to the electrodes forces the lithium ions to migrate from the cathode (the positive electrode) to the anode (the negative electrode), where they embed in the porous cathode graphite material by *intercalation*.[10] During discharging, lithium ions take the reverse path through the separator diaphragm back to the cathode; electric energy is therefore provided to the external circuit connected to the two electrodes.

In this section, we will describe the main characteristics of the following batteries per cathode technology:

- lithium oxide (cobalt/manganese)
- lithium iron phosphate

As a side note, lithium could be considered as a scarce resource, hence a large penetration of such batteries in the market could pose oligopoly threats to the economy with increasing costs, political tensions and eventually the need to find substitutes in the medium term.

---

[9] *Polyanions* are molecules with negative charges positioned in different sites across the molecule structure.

[10] *Intercalation* is the inclusion of a different molecule into a pattern of other different molecules. The inclusion process is usually reversible.

Lithium Oxides

Lithium oxide batteries are built with the positive electrode made from oxides of cobalt or manganese. Cobalt, symbol Co, is a metal with atomic number 27, and common oxidation states of +2 and +3. Manganese, symbol Mg, is a metal with atomic number 25, similar to iron and with common oxidation states +2, +3, +4, +6, and +7.

The reaction that takes place at the positive electrode of a $LiCoO_2$ (with x as the number of moles and the direction of charging going from the right to the left):

$$LiCoO_2 \leftrightarrow Li_{1-x}CoO_2 + x\,Li^+ + x\,e^- \tag{1.48}$$

At the negative electrode, the inclusion process occurs as per:

$$x\,Li^+ + x\,e^- + x\,C_6 \leftrightarrow x\,LiC_6 \tag{1.49}$$

For a $LiMg_2O_4$, the half reactions that occur at the electrodes are the following:

$$Li \rightarrow Li^+ + e^- \tag{1.50}$$

$$MnO_2 + Li^+ + e^- \rightarrow LiMnO_2 \tag{1.51}$$

and the overall reaction is:

$$Li + MnO_2 \rightarrow LiMnO_2 \tag{1.52}$$

Normally, each cell yields a voltage of 3–4 V, which is higher than other types of batteries and makes this device lighter, more compact and more suitable for nonstationary applications. The energy density normally falls in the range of 80–180 Wh/kg.

Cell construction can be cylindrical or prismatic, depending on specific application. Normally, a cylindrical form factor is capable of storing a higher specific energy, but the prismatic shape allows for optimized construction for large scale application. Prismatic form factor also is more suitable for better thermal management, thanks to a better surface-to-volume ratio that facilitates heat exchange.

Shepherd's equations for discharge and charge is similar to the ones described in previous sections. The exponential region of the charge and discharge curves do not show the hysteresis of lead-acid and nickel-based technologies; this is why the exponential factor is the same in discharge and charge equations. As for the previous models, the parameters are extracted from the characteristic curves from experimental setups.

Discharging is modeled with:

$$U_{dis} = E_0 - Ri - K\frac{Q}{Q - \int i\,dt}\left(\int i\,dt + i^*\right) + A\exp\left(-B\int i\,dt\right) \tag{1.53}$$

where the parameters are the ones already discussed in Sec. 1.8.2.

During charging, lead-acid and Li-ion batteries share the same EOC behavior with the voltage the increases rapidly when nearing SOC of 100%. Charge is therefore modeled with the following equation:

$$U_{ch} = E_0 - Ri - K\frac{Q}{|\int i\,dt| - 0.1 \times Q}i^* - K\frac{Q}{Q - \int i\,dt}\int i\,dt + A\exp\left(-B\int i\,dt\right) \tag{1.54}$$

Table 1.5 provides typical value of the main parameters of lithium oxide batteries.

Specific energy (Wh/kg)	100–200
DOD (%)	80–90
RT efficiency (%)	85–95
Typical cycle number	> 3000
Self-discharge (% monthly)	2–8
Response time	< 1 s

TABLE 1.5 Lithium Oxide Battery Main Data

Lithium-Iron-Phosphate

The *lithium-iron-phosphate battery* is a LIB where cathode is made of $LiFePo_4$ coated with carbon and aluminum (or other materials, like zirconium or lithium pyrophosphate) to improve electrical conductivity [8].

Main characteristics of this technology, if compared with $LiCoO_2$ or $LiMn_2O_4$, is the higher number of cycles, the smaller idle capacity losses, good power capability, and safer and better environment-friendly operations than other electrochemical battery technologies, at a cost of lower specific energy. Another advantage of the technology is the very stable discharge voltage characteristic, which remains close to its nominal value (typically 3.2 V) for nearly all the discharge curve.

Applications of $LiFePo_4$ occur mostly in UPS electronics, solar lighting, consumer electronics, and electric vehicles.

Table 1.5 provides typical value of the main parameters of lithium-iron-phosphate batteries.

1.8.5 Zinc-Air

Zinc, symbol Zn and atomic number 30, is an element belonging to the transition metal class, is the 24th most abundant element in Earth's crust, and has only one common oxidation state of +2. *Zn-air batteries* are a kind of metal-air batteries, where a zinc electrode is the anode and an air electrode is the cathode, with KOH as the electrolyte. The main advantages of this battery are the very high specific energy and their inexpensive and environmentally friendly components [9, 10]. The high energy density of zinc-air batteries comes from the fact that Zn-air technology does not need material for the cathode, since atmospheric air itself is used as the cathode, therefore reducing the total volume of the battery.

The reaction taking place at the cathode is, during discharge [11]:

$$O_2 + 2 H_2O + 4 e^- \rightarrow 4 OH^- \tag{1.55}$$

Specific energy (Wh/kg)	90–120
DOD (%)	80–90
RT efficiency (%)	90
Typical cycle number	> 3000
Self-discharge (% monthly)	2–8
Response time	< 1 s

TABLE 1.6 Lithium-Iron-Phosphate Battery Main Data

Specific energy (Wh/kg)	> 150
DOD (%)	80
RT efficiency (%)	75–85
Typical cycle number	> 5000
Self-discharge (% monthly)	15 (near 0 if sealed)
Response time	< 1 s

TABLE 1.7 Zinc-Air Battery Main Data

At the anode:

$$2\,Zn \rightarrow 2\,Zn^{2+} + 4\,e^- \tag{1.56}$$

$$2\,Zn^{2+} + 4\,OH^- \rightarrow 2\,Zn(OH)_2 \tag{1.57}$$

$$2\,Zn(OH)_2 \rightarrow 2\,ZnO + 2\,H_2O \tag{1.58}$$

The overall reaction is:

$$2\,Zn + O_2 \rightarrow 2\,ZnO \tag{1.59}$$

Since cathode is atmospheric air, no changes in the cathode material occur; as a consequence, during discharge, the voltage remains stable until battery is depleted.

Construction of zinc-air batteries must balance the air inflow but at the same time avoiding losses in the water content of the electrolyte. To achieve this, the hydrophobic properties of teflon are taken advantage of in cathode membranes to reduce dehydration during operations. Furthermore, low temperatures might also affect battery capacity, that could be reduced in cold operating conditions; manufacturer's datasheets show a temperature range of 10–50°C.

Applications for zinc-air batteries are in test for electric mobility and energy grid storage, with a system of 4 MWh, 1 MW currently in deployment. Since zinc is a much more common element than lithium, its future use in next electric vehicles might not be hampered by possible oligopolistic behavior of lithium-producing countries.

Table 1.7 provides information on the main Zn-air battery parameters.

1.8.6 Molten Salts

Molten salt batteries can be considered a solid state battery technology, but their operations mode is actually occurring at very high temperatures where salts melt and act more as liquids rather than solid batteries.

Sodium, symbol Na and atomic number 11, is an element belonging to the alkali metal class, is a very reactive element and ranks between the most abundant element that can be found on earth. Solid at ambient temperatures, its melting point is 98°C. It is used as the key element for ion exchange in molten salt batteries in compounds like:

- sodium-sulfur
- sodium-nickel-chloride

Operating at very high temperatures, molten salts technology needs a particularly effective thermal insulation to reduce energy losses in the form of heat, and internal temperature regulation systems to maintain cell temperatures within the correct temperature range for optimal functioning.

Sodium-sulfur

Sulfur, symbol S and atomic number 16, is an element belonging to the nonmetal class, capable of reacting either as an oxidant or as a reducing agent. It is used in combination with sodium in *sodium-sulfur batteries* (NaS). Operating temperatures for NaS batteries range between 300 to 350°C, where sodium and sulfur (having a melting point of 115°C, are in their liquid phase.

In the NaS construction, the sodium chamber acts as the anode, the negative electrode, to which Na provides electrons during the discharge phase. A solid electrolyte made of beta-alumina separates the sodium electrode from the sulfur electrode. Beta-alumina is permeable to sodium ions (Na^+) but not for electrons, avoiding spontaneous self-discharge when the battery is in idle mode. The sulfur cathode acts, during discharge, as the acceptor of the electrons provided by Na to the external circuit, and sulfur reacts with electrons and sodium ions to form sodium sulfides [12, 13]. During charging, the chemical reactions are reversed. The basic chemical reaction is the following:

$$2\,Na + x\,S \leftrightarrow Na_2S_x \tag{1.60}$$

where at the negative electrode the reaction is, during discharge:

$$2\,Na \rightarrow 2\,Na^+ + 2e^- \tag{1.61}$$

and at the positive electrode, during discharge:

$$x\,S + 2\,e^- \rightarrow S_x^{-2} \tag{1.62}$$

where x is comprised between 3 and 5. During charging, the reactions are reversed.

NaS temperature profile changes during battery operations. During discharge, the internal resistance causes the cell to raise its internal temperature. During charge, internal heat generation is compensated by internal reaction heat absorption, and internal temperature tend to lower or remain constant. In idle mode, internal heat is lost through the interface with the environment and internal temperature lowers up to a certain threshold, when an ancillary heat source must be employed to maintain the correct operating temperature range to keep the salts from solidifying. The cell internal resistance decreases if internal temperature increase; temperature therefore has an effect on NaS battery performance changing, for instance, its efficiency and peak power output.

Since the compounds of sodium with sulfur are highly corrosive, special care is to be taken in the design and manufacture of these batteries. This, in combination with the high operating temperatures, poses pressure on cost reduction of molten salt batteries.

NaS battery technology has been demonstrated at over 190 sites in Japan. More than 270 MW of stored energy suitable for 6 h of daily peak shaving have been installed. The largest NaS installation is a 34 MW, 245 MWh unit for wind stabilization in Northern Japan. The use as batteries for nonstationary applications has not been successful due to inherent risks of self-combustion, although mitigated by recent technological advances, corrosion, and high-temperature operations.

Table 1.8 provides the main parameter data for NaS batteries.

Specific energy (Wh/kg)	100
DOD (%)	80–85
RT efficiency (%)	75–85
Typical cycle number	2500–4500
Self-discharge (% monthly)	0 (but heat losses)
Response time	< 1 s

TABLE 1.8   NaS Battery Main Data

Sodium-nickel-chloride

The *sodium-nickel-chloride battery* (NaNiCl) is a molten salt battery similar to NaS batteries [14, 15]. The cut-out schematic of a battery is shown in Fig. 1.9.

The cathode is made of a mixture of nickel powder and $NaAlCl_4$ (sodium-chloroaluminate) inside a beta-alumina tube. A metal current collector pole is fitted inside the beta-alumina tube. $NaAlCl_4$ is solid at ambient temperature, therefore, cell optimal operations occur in the range of 270–350°C. In this temperature interval, the beta-alumina electrolyte offers the least resistance to the flow of ions, and $NaAlCl_4$ melts[11] becoming conductive for sodium ions and permitting their flow inside the cell. Thermal management is therefore fundamental for correct functioning of the system, but poses technological challenges that increase the complexity of the cell and the overall system. The anode is the external case that contains the cathode, the beta-alumina electrolyte, and the sodium located between the beta-alumina tube ad the external case.

The reaction at the positive electrode is:

$$NiCl_2 + 2\,Na^+ 2\,e^- \rightarrow Ni + 2\,NaCl \tag{1.63}$$

The reaction at the negative electrode is:

$$Na \rightarrow Na^+ + e^- \tag{1.64}$$

The overall cell reaction is the following:

$$Ni + 2\,NaCl \leftrightarrow Na + NiCl_2 \tag{1.65}$$

One advantage of NaNiCl technology is that, in case of a cell failure, that cell resistance remains similar to that of a nonfaulty cell resistance, meaning that a series of NaNiCl cells can continue to operate even in case of multiple cell failures. The system is therefore very robust. For instance, a 557 V unit of 18 kWh can be achieved by a single connection of 216 NaNiCl cells. The limit to the number of cells that can enter bypass mode due to failure is determined by the reduction of energy content that is accepted by the designers of the system. If the number of faulty cells goes over a certain threshold, a maintenance intervention can substitute the nonworking modules in a rather easy and quick way, whereas the battery will not stop on its own due to cell faults. To cater for a reduction in capacity, the battery-management system has the capability to adjust the charging voltage according to the overall cell series open circuit voltage, resulting in an uninterrupted operating mode even in case of multiple cell failures.

---

[11] $NaAlCl_4$ liquefies at its melting point of 154°C.

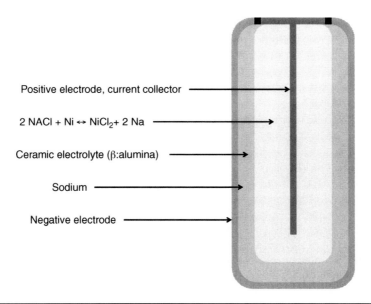

Positive electrode, current collector

2 NACl + Ni ↔ NiCl$_2$+ 2 Na

Ceramic electrolyte (β:alumina)

Sodium

Negative electrode

FIGURE 1.9 Schematic of a NaNiCl molten salt battery.

This technology has a 100% coulombic efficiency with no self discharge, but thermal management is needed to maintain the temperature of at least 270°C; the power needed per cell is in the range of 80 W for a 20 kWh cell, which is an energy loss which must be accounted for during design.

The battery-management system periodically needs to top up the charge of the cell in order to re-set the electronic coulometers to maintain accuracy of capacity measurements. This procedure imposes a periodical full charging that can cause some constraints in battery operations. The battery-management system also takes care of thermal management, trying to maintain operating conditions within the optimal temperature range.

NaNiCl batteries are used in vehicles, especially in public transportation for full-electric or hybrid buses, telecommunication stations, and in maritime applications. Stationary energy storage coupling PV power plant and NaNiCl storage has been developed in a number of countries. One example is a 1.2 MW$_p$ PV plant in northern Italy with a NaNICl ESS of 280 kWh, 390 kW peak power, used for energy injection smoothing. A similar applications has been deployed in the Maldives to provide energy time shifting and smoothing from a solar PV plant, with an ESS of 1.2 MWh, 400 kW.

Table 1.9 provides the main data about NaNiCl batteries.

Specific energy (Wh/kg)	100
DOD (%)	80
RT efficiency (%)	80–90
Typical cycle number	4500
Self-discharge (% monthly)	0 (but heat losses)
Response time	< 1 s

TABLE 1.9 Sodium-Nickel-Chloride Battery Main Data

Specific energy (Wh/kg)	5–15
DOD (%)	100
RT efficiency (%)	75–90
Typical cycle number	3000
Self-discharge (% monthly)	Not available
Response time	< 1 s

TABLE 1.10   AHI Battery Main Data

1.8.7   Aqueous Ions

Aqueous ion (AHI) batteries have been designed as a technology aiming at reducing costs and environmental impact of electrochemical storage [16]. The focus have been put on understanding which are the material that have the best trade-off between market availability, bulk supply, stability at ambient temperature and pressure, lowest possible toxicity, and aggressiveness on other materials. Therefore salt water, carbon, manganese, cotton, and stainless steel have been chosen as the best materials between a wide range of other elements, chemical species, and materials capable of creating cells with electric potentials sufficient to permit the flow of electrons, and polypropylene for the packaging of the stacks.

The electrolyte in AHI batteries is an aqueous solution of sodium sulfate Na_2SO_4 which is inert, low cost, and completely dissociated in water. For electrochemical cell design, the choice of the electrolyte is very important in creating the best conditions for ion transport. This electrolyte used in AHI has both a low resistance, which helps in improving overall battery efficiency, and a low or neutral pH that minimizes the risks of corrosion and aging of the electrodes. Since the electrolyte is basically salt water, the membrane that separates the two half-cells is just made out of simple synthetic cotton, a very low cost material.

The cathode is a spinel[12] λ-MnO_2 obtained from the compound $LiMn_2O_4$, a manganese oxide that is stable in aqueous solutions and therefore does not require surface coating that is normal on electrodes immersed in organic solvent electrolytes.

The anode material is made of *activated carbons* (AC), a type of carbon compound with a very high porosity and therefore characterized by a large specific surface area[13] up to 2800 m^2/g. Carbon electrodes do not undergo corrosion in aqueous environments and carbon have no environmental impact.

The cells are packaged in polypropylene cases, where current collection is performed through graphite sheeting and welded stainless steel tabs to make the electrical connections of the stacks.

The mechanisms of charge exchange are based on sodium intercalation at the cathode, whereby at the anode the mechanism is not yet well understood. One possible hypothesis is that a pseudocapacitive hydrogen effect is responsible for charge exchange in the AC surface; in aqueous solutions, hydrogen can evolve and can be stored in the AC pores where it can recombine with OH^- groups. New developments of the technology are addressing new anode and electrolyte composition for increased cell efficiency.

The main characteristics of AHI batteries are summed up in Table 1.10.

---

[12] A *spinel* is a class of minerals that crystallize in cubic, or isometric, form.

[13] The *specific surface area* is the total surface area per unit mass.

1.9   Redox Flow Batteries

Redox flow batteries (RFB) are electrochemical ESS where two electrolyte are pumped from their own reservoirs into an electrochemical cell (the power block) where their oxidation states change according to whether the system is in charging or in discharging mode. When an electrolyte exits the power block, it is pumped to another reservoir that collects the depleted, or charged, electrolyte with its changed oxidation state. Eventually, since electrolytes are stored in reservoirs, a flow battery can be easily and quickly replenished with charged electrolyte without the need to perform a charging cycle. Normally though, charging energy comes from the feeder line of the distribution grid or the REN source to which the flow battery is connected to (Fig. 1.10).

The electrolyte in the positive reservoir is also called *catholyte*, while the one in the negative is called *anolyte*. In RFB, the electrochemical reactions occur in the power block, a stack of bipolar electrodes where the electrolyte flows, separated by membranes.

An electrical equivalent model for RFB is given in Fig. 1.11. In this model, the stack voltage is considered as a voltage generator function of SOC. Losses are estimated with the internal resistances R_{int} and $R_{resistive}$ and the external losses due to the energy used for the circulation pumps. The model also considers the dynamic behavior in the short term with a capacitance $C_{electrodes}$ that replicates capacitive effects at the electrode interfaces.

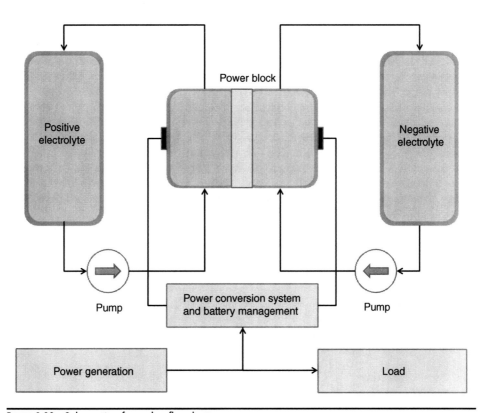

FIGURE 1.10   Schematic of a redox flow battery.

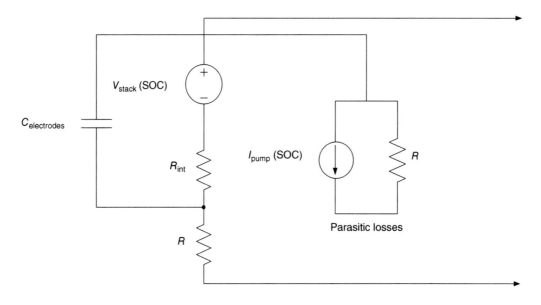

FIGURE 1.11 Electric model for a VSB (modified from [18]).

As already stated, RFB can have a very high energy content depending on how large are the reservoirs for the electrolyte. Energy content is therefore virtually unlimited; by replenishing the electrolyte in the reservoirs, the flow battery can run over an indefinite time range before needing recharging.

Differently from solid-state electrochemical batteries, energy and power can be considered completely decoupled, where energy depends on the volume of the ionic solution, and power by the electrochemical cell where the redox reactions occur. Another advantage of RFB is that full DOD of 100% can in theory be achieved and that, if left uncharged even for long periods, the battery is not going to suffer damages in its core components. Normally, no toxic materials are used in the electrolytes, making RFB a potentially safe-storage technology with ease of integration in many storage applications.

In RFB, energy density tends to be rather low, and economically viable operations could be reduced to energy-intensive applications where long charging and discharging hours are needed. Due to the systems that control and pump the liquids into the power block, system complexity tends to be higher than other energy storage technologies; RFB therefore need a certain amount of mechanical maintenance that increases, as a consequence, the operating expenses of the overall system.

A number of different types of flow batteries have been developed, depending on the chemical species employed in the electrolytes. In the next sections, three types of RFB will be discussed, based on the following reactants:

- vanadium
- zinc-bromide
- iron-chromium

1.9.1   Vanadium Flow Battery

The *vanadium redox battery* (VRB) employs vanadium in different oxidation states to store electric energy in the form of chemical potential energy [19]. Vanadium is a transition metal with atomic number 23 and found in nature only in combined forms. Vanadium is capable of forming solutions with four different oxidation states, which is a property shared only by other few elements in the periodic table. The VRB consists of a stack of cells where the reactions take place, two electrolyte tanks with solutions of vanadium salts in sulfuric acid and pumps to flow the electrolytes in the battery stacks. Stacks are separated by membranes permeable to protons.

The electron exchange at the electrodes occurs according to the following reactions, in which charging is the reaction occurring from left to right, and discharging in the opposite direction. At the positive electrode (cathode) the reaction is:

$$VO^{2+} + H_2O \leftrightarrow VO_2^+ + 2H^+ + e^- \tag{1.66}$$

where VO^{2+} is the vanadium oxide ion of V^{4+}, and VO_2^+ is the oxide ion for V^{5+}. At the negative electrode (anode):

$$V^{3+} + e^- \leftrightarrow V_2^+ \tag{1.67}$$

The overall reaction is:

$$V^{3+} + VO^{2+} + H_2O \leftrightarrow V_2^+ + VO_2^+ + 2H^+ \tag{1.68}$$

Protons pass through the proton exchange membranes (PEM, normally made of trademarked materials like nafion or selenium CVM) and ensure electroneutrality of the system closing the electrical circuit. Table 1.11 summarizes the redox reactions of vanadium during charge and discharge.

The VRB full cycle efficiency is in the range of 70–85%, with more than 10,000 charging–discharging cycles at temperatures of 10–35°C. Normally, the energy density is in the range of 40–70 Wh/kg depending on technology, but is reduced when taking into account all sub-systems that complete the overall storage system. A VRB response time is very short, usually in the range of millisecond. Also, the battery allows overloads of as much as 400% for 10 s and its capacity can be virtually unlimited by simply refilling the tanks. Furthermore, VRB show no memory effects or display any sign of damage in case the two different electrolyte liquids mix with each other. However, the system is fairly complex and needs a large volume to achieve acceptable storage standard, thus resulting in its low energy-to-volume ratio. If a higher volumetric energy density can

During discharging, at cathode: V^{5+} reduces to V^{4+}	at anode: V^{2+} oxidizes to V^{3+}
During charging, at cathode: V^{4+} oxidizes to V^{5+}	at anode: V^{3+} reduces to V^{2+}

TABLE 1.11 Schematic of Vanadium Oxidization State Changes Reactions During Charge and Discharge

be achieved with future developments, the refill capacity of the VRB can help widen its nonstationary applications: for example, electric vehicles equipped with VRB in principle could be replenished at the fuel distribution stations just as quickly as traditional vehicles.

The typical open-circuit voltage of one cell is 1.41 V at a temperature of 25°C. Stack voltage depends on a number of parameters that change during VRB operations: apart from charge or discharge current, temperature, and electrolyte flow rates, cell voltage is a function of vanadium and proton concentrations. The battery single cell voltage is directly related to the SOC of the battery and can be described by the modified Nernst's equation [20]:

$$U_{cell} = U_{eq} + 2\frac{RT}{F}\ln\left(\frac{SOC}{1 - SOC}\right) \tag{1.69}$$

where V_{eq} is the cell potential at SOC of 50%.

Losses as a function of time t are given by the following [17]:

$$U_{losses}(t) = \eta_{act}(t) - \eta_{conc}(t) - \eta_{ohm}(t) - \eta_{ion}(t) \tag{1.70}$$

where η_{act} is the loss due to inertia in start of reaction between chemical species, η_{conc} is the loss associated with differences in the spatial concentration of reactants in the electrolyte volume, η_{ohm} and η_{ion} are the losses due to resistance in the passage of electrons and ions in the bipolar plate electrodes and membrane interfaces. Since these losses are difficult to be correctly estimated one by one, an equivalent resistance is therefore introduced to cater for all energy losses caused by electrochemical and ohmic phenomena. This equivalent resistance is then multiplied by the current flowing in the stack at time t to give the equivalent overpotential U_{losses} as a function of time.

Stack voltage during charge and discharge as a function of time and currents are given in Fig. 1.12 for a standard VRB battery.

Other losses can occur due to cross contamination of the electrolytes from imperfect functioning of the PEM which do not completely prevent vanadium ions to pass through

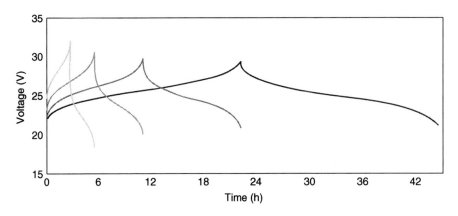

FIGURE 1.12 Stack voltage during charge and discharge at different currents (modified from [17]).

Specific energy (Wh/kg)	20–70
DOD (%)	100
RT efficiency (%)	70–85
Typical cycle number	> 5000
Self-discharge (% monthly)	0
Response time	ms

TABLE 1.12 Vanadium Flow Battery Main Data

the PEM itself. This contamination causes over time a reduction in the concentration of V^{+5} and V^{+2} as per the following:

$$V^{5+} + V^{3+} + 2\,V^{2+} \Rightarrow 4\,V^{3+} \tag{1.71}$$

and:

$$V^{4+} + V^{2+} + 2\,V^{5+} \Rightarrow 4\,V^{4+} \tag{1.72}$$

This causes a performance decay of the VRB which must be taken into account and can lead to full electrolyte substitution after a period of time that has an effect on overall economic viability of the storage system.

Main characteristics of VRB are summarized in Table 1.12.

Many are already the installations of VRB, and RFB in general, if coupled with REN power plants. VRB are, for instance, used to control the injection profile of solar PV output to improve grid stability.

An example of integration between VRB, PV, and wind energy is given in Fig. 1.13, to ensure grid stabilization during solar radiation or wind speed changes. The PV array is connected to the DC bus through a DC/DC boost converter with an MPPT controller;

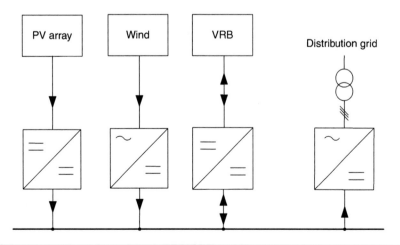

FIGURE 1.13 Example of integration between VRB, PV, and wind energy.

the VRB is connected to the DC bus by a bidirectional DC/DC converter to regulate charge and discharge cycles; finally, an inverter converts DC to AC current and manages the injection of electric energy in the distribution grid. A buck-boost converter is used to control and maximize the power output from wind; the converter's output power is maintained constant to facilitate charging operations of the RFB, which discharges when wind power injection goes below a target power threshold.

1.9.2 Zinc-Bromide Flow Battery

Zinc-bromide ($ZnBr_2$) is an inorganic salt formed by the reaction between zinc oxide ZnO with hydrobromic acid HBr as in:

$$ZnO + 2\,HBr + H_2O \rightarrow ZnBr_2 + 2H_2O \tag{1.73}$$

The *zinc-bromide flow battery* (ZBB) uses the same electrolyte $ZnBr_2$ in two reservoirs, when one reservoir is used for the positive electrode reaction, and the other for the negative electrode reactions. As usual in RFB, the power block consists of a series of cells in a stack configuration, with each cell having a nominal voltage around 1.83 V. Commercial ZBB range in capacity from 50 kWh for a single module, up to over 500 kWh when connected in multiple modules. ZBB appear as having an unlimited number of cycles and DOD of 100%, although some experimental data suggest that phenomena of corrosion can introduce performance decays, and minimum SOC of 20% should be probably respected during operations.

Energy density is in the range of 65–85 Wh/kg, operating temperatures are from −10 to +45°C, and efficiency is around 75–80% depending on charge and discharge conditions. The case and package can be made entirely of recyclable materials (like plastic) to reduce costs and provide ease of installation and end-of-life disposal. The low-toxicity electrolytes add safety to the list of advantages of ZBB technology. It is possible that some battery management procedures be put in place to avoid deposits of zinc on the electrode plates, like periodical full discharge after a number of days or shunting terminals after a number of charge–discharge cycles. Advances in technology are addressing these issues.

The reactions in the zinc-bromide ESS are:

$$Zn \leftrightarrow Zn^{2+} + 2e^- \tag{1.74}$$

$$Br_2 + 2e^- \leftrightarrow 2Br^- \tag{1.75}$$

with the following overall cell reaction:

$$Zn + Br_2 \leftrightarrow 2Br^- + Zn^{2+} \tag{1.76}$$

When fully discharged, the electrolyte is a homogeneous aqueous solution of $ZnBr_2$. During charging, the zinc ions Zn^{2+} in solution are reduced to metallic zinc Zn on the anode; bromide ions Br cross the porous membrane toward the cathode where are oxidized into molecular bromine at the electrode within the aqueous solution. The anolyte and catholyte gradually develop different compositions (with typical changes in the color of the electrolytes). Elemental bromide Br_2 produced in the cathode half-cells combines

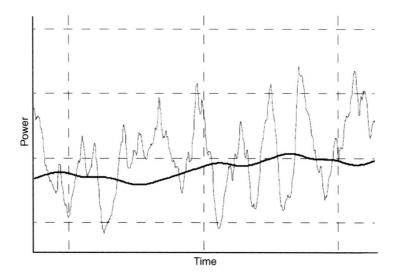

FIGURE 1.14 Power smoothing and ramp control with ZBB for wind power injection (modified from [21]).

with an oil and forms a polybromide complex in the catholyte which separates from the catholyte aqueous phase as a higher density liquid and deposits at the bottom of the catholyte reservoir. The energy is therefore stored in the system as the chemical potential energy of the zinc metal plated across the anode and the polybromide complex in the catholyte reservoir. During discharge, these processes are reversed. Upon full discharge, both the anolyte and catholyte tanks are returned to a homogeneous aqueous solution of zinc-bromide.

The curve that relates voltage to current as a function of time is similar to those of other RFB technologies. As long as there is enough energy content in the electrolytes, the voltage slightly increases during charge and slightly decreases during discharge, with currents that tend to be constant during the process. When the energy content in the electrolyte is not sufficient to provide energy and power to the load, the voltage quickly drops to zero.

Applications for ZBB are analogous to other RFB technologies. A 50 MWh ZBB energy storage is simulated via software modeling to provide power shifting, ramp control, and power smoothing for a grid-connected wind farm of 150 MW [21]. Eighty percent of the total energy storage capacity is reserved for power shifting, and 20% for quicker intervention times in the range of a few minutes. Since the energy converted by a wind turbine is proportional to the cube of the wind speed, small differences in wind speed can cause large imbalances of energy injection, with potential disruptions to the distribution line. The simulation results for the power smoothing application are shown in Fig. 1.14, where such imbalances are smoothed out by the use of the ZBB.

Typical characteristics of ZBB are summarized in Table 1.13.

Specific energy (Wh/kg)	65–85
DOD (%)	100
RT efficiency (%)	75–80
Typical cycle number	> 5000
Self-discharge (% monthly)	0
Response time	ms

TABLE 1.13   Zinc-Bromide Battery Main Data

1.9.3   Iron-Chromium Flow Battery

The *iron-chromium battery* (ICB) is another type of RFB where the anolyte is chromium in its two oxidation states, Cr^{2+} and Cr^{3+}, and the catholyte is iron with Fe^{2+} and Fe^{3+}. Standard cell voltage is in the range of 1.7–3.3 V, energy densities at 50–85 Wh/kg, and efficiency in the 70–80% range.

A construction advantage for ICB lies in the use of low cost porous membrane, since electrolyte can sustain a certain degree of contamination with no significant performance losses.

Commercial modules are manufactured for medium to large scale grid-connected applications, with power starting from 125 kW and energy content of 0.5 MWh (4 h storage), up to 50 MW and 1 GWh. Due to low energy densities, footprint tend to be large and in the range of 35–40 m^2. Efficiency also tends to be higher at temperatures of 40–60°C rather than at ambient temperature. Safety, sufficiently good efficiency and low cost can make this technology interesting for many industrial and utility scale applications, like smoothing or profiling of REN, demand charge management, distribution and transmission quality of service in utility-scale plants. Reactants are also easy to reuse, making decommissioning of ICB easier and cost-effective.

A medium-sized REN ICB plant is currently in operations to store energy from a 150 kW solar system and power a 260 kW irrigation pump in an agricultural setting.

Typical characteristics of iron-chromium batteries are summarized in Table 1.14.

Specific energy (Wh/kg)	50–85
DOD (%)	100
RT efficiency (%)	70–80
Typical cycle number	> 5000
Self-discharge (% monthly)	0
Response time	ms

TABLE 1.14   Iron-Chromium Battery Main Data

References

1. IUPAC Compendium of Chemical Terminology — The Gold Book, http://goldbook.iupac.org/.
2. Kiehne H. A. (2003). Battery Technology Handbook. CRC Press, New York, USA.

3. Chen M., Rincón-Mora G. A. (2006). Accurate electrical battery model capable of predicting runtime and I–V performance. IEEE Transactions on Energy Conversion 21(2):504–511.

4. Lambert T., Gilman P., Lilienthal P. (2006). Micropower System Modelling with Homer, in Integration of Alternative Sources of Energy, by Felix A. Farret and M. Godoy Simões. Wiley Hoboken, New Jersey, USA, pp. 379–418.

5. Jongerden M. R., Haverkort B. R. (2009). Which battery model to use? Software, IET 3(6):445–457.

6. Shepherd C. M. (1965). Design of Primary and Secondary Cells — Part 2. An equation describing battery discharge. Journal of Electrochemical Society 112:657–664.

7. Hussein A. A-H., Batarseh I. (2011). An overview of generic battery models. Power and Energy Society General Meeting, 2011 IEEE, 24–29 July, pp. 1–6.

8. Padhi A. K., Nanjundaswamy K. S., Goodenough J. B. (1997). Phospho-olivines as positive-electrode materials for rechargeable lithium batteries. Journal of The Electrochemical Society 144:1188–1194.

9. Deiss E., Holzer F., Haas O. (2002). Modeling of an electrically rechargeable alkaline Zn-air battery. Electrochimica Acta 47:3995–4010.

10. Caramia V., Bozzini B. (2014). Materials science aspects of zinc-air batteries: a review. *Materials for Renewable and Sustainable Energy* 3:28.

11. Mohamad A. A. (2006). Zn/gelled 6 M KOH/O_2 zinc-air battery. Journal of Power Sources 159:752–757.

12. Kamibayashi M., Tanaka K. (2001). Recent sodium sulfur battery applications. Proceedings of the IEEE PES Transmission and Distribution Conference and Exposition, USA, 28 Oct.–2 Nov. 2001, vol. 2, pp. 1169–1173.

13. Hussien Z. F., Cheung L. W. et al. (2007). Modeling of sodium sulfur battery for power system applications. Elektrika 9(2):66–72.

14. Sudworth J. L. (2001). The sodiumnickel chloride (ZEBRA) battery. Journal of Power Sources 100:149–163.

15. Dustmann C. H. (2004). Advances in ZEBRA batteries. Journal of Power Sources 127:85–92.

16. Whitacre J. F., Wiley T., Shanbhag S., Wenzhuo Y., Mohamed A., Chun S. E., Weber E., et al. (2012). An aqueous electrolyte, sodium ion functional, large format energy storage device for stationary applications. Journal of Power Sources 213:255–264.

17. Blanc C., Rufer A. (2008). Multiphysics and energetic modelling of a vanadium redox flow battery. IEEE International Conference on Sustainable Energy Technologies, Singapore, 24–27 Nov. 2008, pp. 696–701.

18. Barote L., Marinescu C., Georgescu M. (2009). VRB modeling for storage in stand-alone wind energy systems. IEEE Bucharest Power Tech Conference, June 28th – July 2nd 2009, Bucharest, Romania, pp. 1–6.

19. Poullikkas A. (2013). A comparative overview of large scale battery systems for electricity storage. Renewable and Sustainable Energy Reviews 27:778–788.

20. Chahwan J., Abbey C., Joos G. (2008). VRB modelling for the Study of Output Terminal Voltages, Internal Losses and Performance." IEEE Electrical Power Conference-EPC'07, 25–27 Oct. 2007, Montreal, Canada, pp. 387–392.

21. Esmaili A., Nasiri A. (2010). Energy storage for short-term and long-term wind energy support. IECON 2010 — 36th Annual Conference on IEEE Industrial Electronics Society, 7–10 Nov., Glendale USA, pp. 3281–3286.

CHAPTER 2

Chemical Storage

Summary. Renewable energy can be converted and stored into chemical energy mainly in gaseous form. Electrical energy is then retrieved by combustion processes in fuel cells (FCs) or turbines. Hydrogen and methane are between the most common options as gases obtained from power-conversion processes. One of the main advantages of chemical energy is the fact that it can be stored for long periods of time when gases are stored in high-capacity tanks, and can be also be used in nonstationary applications as a fuel for quick refilling of electric vehicles with long driving ranges.

2.1 Introduction

Power-to-gas (PTG) is the process that converts electrical energy into chemical energy by means of electrochemical or chemical reactions. Normally, the chemical species that are obtained are in gaseous form, most often hydrogen or methane. Gases can be relatively easily stored in tanks for later use.

Conversely, the term *gas-to-power* entails the conversion of the energy stored in chemical form to electrical energy by means of combustion of the gas in FCs, where chemical energy is directly converted to electrical energy, or thermal cycles in turbines, where chemical energy is converted first in mechanical energy and then in electrical energy.

This way of storing electrical energy has a big advantage over other storage technologies: the capability to store potentially enormous amounts of energy for a very long period of time, since practically no self-discharging takes place when the gas is stored in tanks. As a consequence, PTG is an alternative storage solution to hydroelectric or compressed air plants. Virtually all large-scale power and energy applications for transmission and distribution (T&D) operators can be implemented with PTG plants. A time-of-use application is not restricted only to intra day operations, but PTG makes seasonal, and strategic, storage viable; the dispatch of large amounts of renewable energy can occur therefore months after REN energy conversion [1]. Furthermore, by injecting gas in the already existing gas transport piping network, the capability to transport energy to different and very far away locations can provide an important additional benefit to a national energy system. If energy conversion from REN happens, for instance, in areas where there is no real need for its full or immediate consumption, the conversion of electrical energy to gas and its injection in the gas grid can provide to far away areas all the energy that is needed. This results in a better integrated overall energy infrastructural

43

system. As an additional use, gas can be employed easily for nonstationary applications, enabling quick refilling and long-range use of electric or natural gas vehicles.

2.2   Hydrogen

Hydrogen has been considered to be used as an energy fuel since when it was first identified as an isolated chemical element by Cavendish in 1766, and received its name (meaning *water-former*) from Lavoisier in 1781; even Jules Verne described hydrogen as the coal of the future in one of his novels. In 1839, Grove conducted the earliest experiments on the use of hydrogen in a prototypical FC; in 1870, Otto used a mixture containing 50% of hydrogen in his experiments for the first internal combustion engine; and in 1923, Haldane successfully produced hydrogen with the energy from a windmill. In 1938, Sikorski recognized the viability of hydrogen as a fuel for transportation as a propeller gas for helicopters [2–6], while nowadays hydrogen is used mostly for aerospace applications.

Hydrogen is an *energy carrier*, or a *secondary energy source*, whereas energy sources like wood, petroleum, and coal are *primary* since they are immediately available for energy uses. Indeed, before its utilization, hydrogen must be obtained in its molecular form (H_2) from other compounds. As an energy carrier, hydrogen can be used in combustion processes directly in thermal cycles (i.e., in turbines), or in FC, where it becomes possible to avoid the performance limitation of the second law of thermodynamics that occurs in direct combustion. As an additional advantage, the reaction of hydrogen with oxygen in a FC yields energy without releasing CO_2, which is the typical by-product of fossil fuel combustion. In terms of environmental impact, hydrogen can be the best alternative to fossil fuels, as it drastically reduces the release of green-house gases and polluting compounds (i.e., sulfur, and fine particles) harmful to the environment and to human health.

Hydrogen embeds a very high gravimetric energy density but a very low volumetric density; this means that transport and storage of large quantities of hydrogen can be complicated and expensive, even with the technologies that have been developed since decades by the chemical industry where hydrogen is commonly produced, stored, and used in many industrial processes.

2.2.1   Properties

Hydrogen (symbol H, atomic number 1 and electron configuration $1s^1$) is a nonmetal element belonging to group I A with atomic weight of 1.00794. At standard temperature and pressure it is a colorless, odorless, and tasteless gas. The hydrogen atom is composed of a proton and an electron, has two oxidation states ($+1$, -1) with an electronegativity of 2.2 on the Pauling scale. Hydrogen's electron is very reactive and easily builds covalent bonds with other hydrogen atoms to reach the stable configuration with diatomic molecular form H_2 [7, 8]. There are two hydrogen isotopes: *deuterium*, with one proton and one neutron and *tritium* with one proton and two neutrons. Every proton in a hydrogen molecule has its own type of spin which results in the existence of two types of H_2 molecules: *ortho-hydrogen*, where the protons of the two atoms have the same spin, and *para-hydrogen*, where the protons of the two atoms have opposite spin. Hydrogen is the most commonly found element in the universe, and one of the main components of stars and interstellar gases, composing a 75% of the Sun's mass. Hydrogen is also one

Molecule	H_2
Phase at STP	Gas
Melting point	14.025 K
Boiling point	20.268 K
Critical point	32.9 K
Molar volume	$11.42 \cdot 10^{-3}$ m^3 mol
Enthalpy of vaporization	0.44936 kJ/mol
Enthalpy of fusion	0.05868 kJ/mol
Density	0.0899 kg/m^3
Density	0.084 kg/Nm^3
Electronegativity	2.2 (Pauling scale)
Thermal conductivity	0.1815 W/(mK)
Ionization energies	1312.06 kJ/mol
Specific heat	14.9 kJ/(kg K)
Lower heating value	110.9–10.1 (MJ/kg–MJ/Nm^3)
Minimum ignition energy	0.02 mJ
Flammability limits (by volume percentage)	4–75%

TABLE 2.1   Physical Properties of Hydrogen

of most abundant elements on Earth and can be found in a wide range of organic and inorganic molecules like water, hydrocarbons, carbohydrates, and amino acids.[1] Isolated hydrogen molecules though are extremely rare to be found on Earth.

Some of the physical properties of hydrogen are summarized in Table 2.1.

2.2.2   Hydrogen Production from Water Electrolysis

Electrolysis is the process that makes use of electrical energy to break chemical compounds into more elementary compounds. A special case is *water electrolysis*, where, in an electrochemical cell, water is separated into hydrogen and oxygen as per the following

---

[1] The *hydrogen bond* is a type of weak electrostatic bond that forms when a partially positive hydrogen atom covalently bonded in a molecule is attracted by another partially electronegative atom equally covalently linked to another molecule. It is described in particular as a dipole–dipole interaction when a hydrogen atom shares a covalent bond with highly electronegative elements like nitrogen, oxygen, and fluorine which attract valence electrons and acquire a partial negative charge, leaving the hydrogen with a partial positive charge. The hydrogen bond is present in both the liquid and the solid states of water. The hydrogen bond keeps the molecules more distant from one another with respect to other types of chemical bonds, which is why ice has a lower density than water. In fact, water molecules float as liquid but form a crystal structure in the solid state as ice. Hydrogen bonds also exists in proteins and nucleic acids and act as one of the main forces to unite the base pairs of the double helical structure of DNA. This bond is fundamental for the equilibrium of our ecosystem. Without it, for example, water would have very different physical properties and the current life forms as we know on this planet would be impossible to exist.

reaction: The overall reaction is therefore:

$$2\,H_2O \rightarrow 2\,H_2 + O_2 \tag{2.1}$$

The *two laws of electrolysis*, or *Faraday's laws* are the following:

1. the quantity of the elements produced during electrolysis is directly proportional to the amount of the electricity passing through the electrolytic cell;
2. with a given quantity of electricity, the amount of the elements produced is proportional to the equivalent weight[2] of the element.

An electrolytic cell comprises of an electrolyte, in the form of an acid (like HCl), a base (a hydroxide like NaOH), or a salt (like NaCl), and two electrodes immersed in the electrolyte. The electrolyte reacts with the solvent (usually water) and splits into positive and negative ions (i.e. H^+, Na^+, OH^-, or Cl^-). By connecting the two electrodes to an outer electric circuit in which an electromotive force is applied, an electron flow is produced through the external circuit and is corresponded by an ion flow in the cell internal electrolyte solution. The cathode is the electrode where reduction half-reactions occur (electrons entering the cell), while the anode is the electrode where the oxidation half-reactions (electrons exiting the cell) take place [9–12].

Industrial electrolyzers consist of more than one electrolytic cell connected in series with metallic separators between two subsequent cells working as a bipolar plate (anode on one side and the cathode on the other). In case of water electrolysis, the efficiency is calculated as the ratio between the chemical energy contained in the yielded hydrogen and the electric power employed to the process. They are categorized according to their technology.

Alkaline Electrolyzers

In an *alkaline electrolyzer* (AE) the electrolytic cells comprise of two electrodes that are immersed in a water solution of potassium hydroxide (KOH), which dissociates int k^+ and OH^- ions. To obtain the electrolysis of water, energy must be provided to the cell in the form of electric current, causing the separation of water into oxygen and hydrogen according to the following reactions.

The OH^- ions are oxidized at the anode releasing diatomic oxygen molecules, water, and electrons as in:

$$4\,OH^- \rightarrow O_2 + 2\,H_2O + 2\,e^- \quad \text{(anode)} \tag{2.2}$$

The oxidation potential ΔV_{rid} of the reaction is -0.40 V. At the cathode, the released electrons are not absorbed by reduction to the metallic potassium of the K^+ ions contained in the liquid solution, since such reaction has a very low reduction potential

---

[2] In chemistry, the *equivalent weight* (or *equivalent mass*) is defined as the quantity of mass of a substance able to supply or consume 1 mole of electrons in a redox reaction, or to generate 1 mole of H^+ ions by dissociation or one mole of OH^- ions in an acid–base reaction. It is calculated as the ratio between the molecular weight of the substance (expressed as g/mol) and its number of moles participating in the reaction. A *mole* is the quantity of a substance that contains the same amount of elementary entities as the number of atoms present in 12 g of C^{12}. Such number is known as the *Avogadro's number*, equal to 6.022×10^{23}.

($\Delta V_{\text{rid}} = -2.92\,\text{V}$). Instead, the reaction that takes place at the cathode is the reduction of water:

$$4\,H_2O + 2\,e^- \rightarrow 2\,H_2 + 4\,OH^- \quad (\text{cathode}) \tag{2.3}$$

The overall reaction is the one given in Eq. (2.1).

This reaction has a reduction potential of -0.83 V, which is still negative but higher than that of the K^+ ions. Therefore, at the cathode the water undergoes reduction (i.e., gains electrons) and releases hydrogen molecules and OH^- ions.

The energy requested by the overall reaction of electrolysis is:

$$\Delta V_{\text{oss}} + \Delta V_{\text{rid}} = -0.40\,\text{V} + (-0.83\,\text{V}) = -1.23\,\text{V} \tag{2.4}$$

which equals the electromotive force required to trigger the two nonspontaneous reactions.

A porous diaphragm, which is permeable to OH^- ions and water but not to H_2 and O_2 gases, permits water and the ion flows while maintaining the separation between the two gases so that they do not mix with each other.

In AE, the anode is made of nickel, the cathode of nickel coated with platinum. The operating temperature is between 70 and 85°C and the electric current density on the electrodes is around 6–10 kA/m^2 with an efficiency ratio between 75 and 85%. AE are designed and built with suitable materials that have to withstand the attack of KOH and avoid its leakages. AE are between the most widespread electrolyzers in the market these days.

Solid Polymer (Polymeric Membrane) Electrolyzers

In *solid polymer electrolyzer* (SPE), or *polymeric membrane electrolyzer* (PME), the electrolyte is a solid polymer, which makes the construction and maintenance simpler and economic. When the polymer is becomes fully impregnated with water, it becomes permeable to ions. Solid construction grants the possibility to operate SPE at pressures up to 4 MPa with temperatures between 80 and 150°C. The cathode of an SPE is made with porous carbon and the anode with porous titanium. Such construction characteristics render the SPE capable of reaching good efficiency and high current densities.

High-Temperature Electrolyzers

High-temperature electrolyzers (HTEs) operates at temperature around 1000°C, need special materials in the solid and noncorrosive electrolyte (built with zirconium and yttrium oxides) that at such high temperature are permeable to ions, but provide very high performances. In HTE, the cathode can be in nickel while the anode in nickel, nickel oxides, and lanthanum. Current densities at the electrodes is normally around 3–5 kA/m^2 at cell voltages in the range of 1.0–1.6 V. Theoretical efficiency is very high, reaching 95%.

Thermodynamics

Assuming that hydrogen and oxygen are ideal gases, that water is incompressible and that its liquid and gaseous phases are separated, then the variations of enthalpy, entropy, and Gibbs' free energy of the electrolysis reaction can be calculated using pure hydrogen, oxygen, and water in standard conditions. As a consequence, the total enthalpy variation of the system during the electrolysis of water is the enthalpy difference between hydrogen, oxygen, and water as in:

$$\Delta H = \Delta H_{H_2} + \frac{1}{2}\Delta H_{O_2} - \Delta H_{H_2O} \qquad (2.5)$$

from which:

$$\Delta H_x = c_{p,x}\left(T - T_{ref}\right) + \Delta H_{f,x}^0 \qquad (2.6)$$

Similarly, the entropy variation ΔS is:

$$\Delta S = \Delta S_{H_2} + \frac{1}{2}\Delta S_{O_2} - \Delta S_{H_2O} \qquad (2.7)$$

where:

$$\Delta S_x = c_{p,x}\,\ln\left(\frac{T}{T_{ref}}\right) - R\,\ln\left(\frac{p}{p_{ref}}\right) + \Delta S_{f,x}^0 \quad \text{for } x = H_2,\ O_2 \qquad (2.8)$$

$$\Delta S_{H_2O} = c_{p,H_2O}\,\ln\left(\frac{T}{T_{ref}}\right) + \Delta S_{f,H_2O}^0 \qquad (2.9)$$

where:

- $c_{p,x}$ is the specific heat of the species x at a constant pressure (for hydrogen molecules it is 28.84 J/mol K, for oxygen molecules 29.37 J/mol K, and for liquid water 75.39 J/mol K)
- ΔH_x is the enthalpy variation of the species x in J/mol
- $\Delta H_{f,x}^0$ is the enthalpy of the formation of the species x in standard conditions (the formation enthalpies of hydrogen and oxygen molecules are both zero by definition)
- p is the pressure measured in Pa (1 atm = 101.325 Pa)
- R is the universal constant of the gas, which equals 8.314 J/mol K
- ΔS_x is the entropy variation of the species x in J/mol K
- $\Delta S_{f,z}^0$ is the entropy of the formation of the species x in standard conditions in J/mol K
- T is the temperature in K

In STC, water electrolysis is not a spontaneous reaction, needing a positive variation in Gibbs' free energy, which in STC is given by $\Delta G_{s,H_2O}^0 = 237$ kJ/mol. ON the contrary, the opposite reaction is spontaneous with a negative free energy of $\Delta G_{f,H_2O}^0 = -237$ kJ/mol. The enthalpy generated by water splitting is 286 kJ/mol and the enthalpy of water formation equals -286 kJ/mol, with $\Delta S_{f,H_2O}^0 = 0.16433$ kJ/mol K.

The work needed for the electrolysis reaction ($\Delta G > 0$) is given by:

$$L_{el} = \Delta G = qE \qquad (2.10)$$

where q is the electron charge transferred in the external circuit of the cell (in C/mol). One mole of water produces a charge of q composed by z moles of electrons as per Faraday's laws:

$$L_{el} = \Delta G = q\,E = z\,F\,E \tag{2.11}$$

where $z = 2$ and F is the Faraday's constant.

In an electrochemical reaction, E is the *reversible cell voltage* U_{rev}, therefore:

$$U_{rev} = \frac{\Delta G}{z\,F} \tag{2.12}$$

Analogously, the *thermoneutral voltage* U_{th} is defined as the electromotive force required by the reaction to occur at a constant temperature:

$$U_{th} = \frac{\Delta H}{z\,F} \tag{2.13}$$

in which $U_{rev} = 1.229$ V and $U_{th} = 1.482$ V in STC.

An increase in temperature would decrease the reversible voltage (at 80°C and 1 bar: $U_{rev} = 1.184$ V) while the thermoneutral voltage would stay almost the same (at 80°C and 1 bar: $U_{th} = 1.473$ V). A rise in the system pressure would cause an increase in the reversible voltage (at 25°C and 30 bar: $U_{rev} = 1.295$ V) while the thermoneutral voltage would remain the same.

Polarizations of the Electrochemical Cell

The efficiency reduction is called *polarization* (or *overpotential, overvoltage*). The phenomena that give rise to polarization involve both the anode and the cathode, depending on the direction of the reaction. Polarization tends to increase, or decrease, the anode voltage where the oxidation reaction takes place and decrease, or increase, the cathode voltage where the reduction reaction occurs. For this reason, they tend to increase the electromotive force needed for electrolysis or lower the output voltage of the FC. Polarization is the final effect of a series of different causes.

For a chemical reaction to occur, the reaction must be capable of overcoming the activation energy, which is the minimum energy required for the reaction to start. *Activation polarization* η_{act} is therefore the minimum voltage required between the electrodes of the cell to initiate the reactions. In an electrochemical reaction this voltage is in the range of 50–100 mV. η_{act} is expressed by the *Tafel's equation* as:

$$\eta_{act} = \frac{R\,T}{\alpha\,z\,F} \log \frac{i_o}{i} \tag{2.14}$$

where α is the coefficient of the *charge transfer coefficient*, i_o is the density of the exchange current and i is the density of the current passing through the electrode surface. The charge transfer coefficient depends on the reaction mechanisms between the electrons and the catalysts and usually acquires a value between one and zero.

Ohmic losses depend on the electrode material resistance to the electron flow, as well as on the electrolyte resistance to the ion flow; *ohmic polarization* can be expressed by the equation:

$$\eta_{ohm} = I\,R \tag{2.15}$$

where I is the current in the cell and R is the total resistance of the cell.

Concentration polarization manifests itself mainly when the current densities are high, and is caused by the slow diffusion of the reactants in the electrolyte that bring about strong concentration gradients, and subsequent voltage changes. From *Fick's first law*,[3] the diffusive transport can be described as:

$$i = \frac{n F D (C_B - C_S)}{\delta} \tag{2.16}$$

in which D is the reactants diffusion coefficient, C_B is their concentration in the electrolyte, C_S is the concentration on the electrode surface, and δ is the thickness of the diffusive layer. When C_S is close to zero, the maximum limit value of i_L can be calculated and it can be reached when the concentration of the reactants at the entry point is too low.

With a few passages from Eq. (2.16):

$$\frac{C_S}{C_B} = 1 - \frac{i}{i_L} \tag{2.17}$$

From the Nernst's equation at equilibrium (hence with no currents inside the cell):

$$E_{i=0} = E_0 + \frac{R T}{n F} \log C_B \tag{2.18}$$

that becomes, for values outside of equilibrium (non-null currents):

$$E = E_0 + \frac{R T}{n F} \log C_S \tag{2.19}$$

From the previous equations, the electrode voltage variation caused by the change of the concentration is finally expressed by:

$$\eta_{conc} = \Delta E = \frac{R T}{n F} \log \left(\frac{C_S}{C_B} \right) = \frac{R T}{n F} \log \left(1 - \frac{i}{i_L} \right) \tag{2.20}$$

The voltage variation occurring at the electrodes depends on the behavior of the same electrodes. The overpotential required to drive the current flow between the electrodes is called *transfer polarization*.

The *reaction polarization* occurs when the chemical reaction in the cell yields new chemical species or changes the equilibrium of the reaction. The variations in the concentrations of the reactants and the products during the cell operation therefore can reduce the conversion efficiency. The synthesis of water, for example, dilutes the solution itself and changes the electrolytes concentration on the electrodes surface.

A temperature increase can improve cell conductivity and reduce the losses related to ohmic polarization. A higher temperature also improves its chemical kinetics and lowers the losses of activation polarization. High-temperature operations cause material decay, like the electrolyte, catalysts, and electrodes. A pressure increase at the cell entry point

---

[3] The *Fick's first law* states that the flux of molecules in a fluid occurs from high-concentration areas to low-concentration regions. The flux J of diffusion is given by: $J = -D \nabla \phi$, where D is the diffusion coefficient that depends on the size of the diffusing molecules, fluid temperature, and fluid viscosity, while ϕ is the three-dimensional concentration of the molecules. The *Fick's second law* gives the change in time of the concentration when the molecules diffuse in a fluid as: $\frac{\partial \phi}{\partial t} = D \nabla^2 \phi$.

also boosts the partial pressure of the reactants improving the solubility of the gas in the electrolyte, with the beneficial effect of reducing concentration polarizations. Excess pressure though impose stress on the device structures potentially reducing system life time. Finally, heating and pressurization systems that can be introduced to improve the operations of the cell have the important trade-off of introducing additional energy consumption, therefore, reducing the efficiency of the overall system.

Mathematical Model

The relationship between current and voltage in an electrolyzer cell is described by mathematical models that are empirical in nature. One possible I–U characteristic curve can be modeled by the following relationship:

$$U_{el} = U_{el,0} + C_{1,el} T_{el} + C_{2,el} \log\left(\frac{I_{el}}{I_{el,0}}\right) + \frac{R_{el} I_{el}}{T_{el}} \tag{2.21}$$

where $U_{el,0}$, $C_{1,el}$, $C_{2,el}$, $I_{el,0}$, and R_{el} are parameters that are obtained from measurements on the cell, T_{el} is the cell temperature, U_{el} is the voltage measured at the electrodes, and I_{el} is the current flowing in the cell (see Fig. 2.1). The first two terms represent the thermoneutral voltage of all the cells connected in series in ideal condition, calculated as the sum of the voltage from each individual cell. The third term includes the losses from the activation polarization while the forth term stands for the ohmic polarization losses, with the losses from the concentration polarization implicitly embedded in the parameters defined by data from experiments.

The operating point of the electrolyzer in any given circumstance when power P_{el} is supplied to the electrolyzer cell is computed by numerical methods applied to the nonlinear system between Eq. (2.21) and $P_{el} = U_{el} I_{el}$.

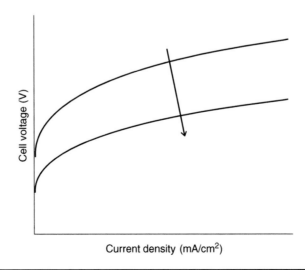

FIGURE 2.1 I–U characteristics in an electrolytic cell at different temperatures.

In accordance with the Faraday's laws, the flow rate of the hydrogen produced is proportional to the electric current passing through the external circuit:

$$\dot{n}_{H_2} = \eta_F \frac{N_c \, I_{el}}{z \, F} \tag{2.22}$$

where $\dot{n}_{H_2}$ is the molar flow rate of the hydrogen produced (in mol/s) and N_c is the number of the electrolytic cells connected in series in the electrolyzer. The molar flow rate of the water supplied and that of the oxygen produced can be described by the following:

$$\dot{n}_{H_2O} = \frac{1}{2} \dot{n}_{O_2} = \dot{n}_{H_2} \tag{2.23}$$

where η_F is the Faraday performance of the electrolyzer, defined as the ratio between the number of moles effectively produced by the electrolyzer and the theoretical maximum obtainable number.

The *electric efficiency* η_e is defined as the ratio between the thermoneutral voltage sum of all the cells in the electrolyzer and the voltage of the electrodes U_{el}:

$$\eta_e = \frac{N_c \, U_{th}}{U_{el}} \tag{2.24}$$

Typical electrolyzer efficiency is in the range of 70%.

2.2.3   Electric Energy Retrieval from Hydrogen Combustion in Fuel Cells

FCs are electrochemical devices where the spontaneous combustion of hydrogen (as the fuel) and oxygen (as the oxidizer) occurs [13, 14]:

$$H_2 + \frac{1}{2} O_2 \rightarrow H_2O + 2\,e^- + heat \tag{2.25}$$

This reaction gives water, electricity, heat, while avoiding any emission of greenhouse gases and pollutants. An FC is actually a combined heat and power system, yielding both electric and thermal energy.

In the FC, hydrogen diffuses at the anode by passing through the electrode pores; by the reaction with the catalyst, hydrogen is adsorbed and ionized, while releasing an electron at the electrode. The reaction is expressed as:

$$H_2 \rightarrow 2[H] \rightarrow 2\,H^+ + 2\,e^- \tag{2.26}$$

in which the square parenthesis indicate the intermediate adsorption process occurring inside the electrode porous structure. The hydrogen ions then pass through the electrolyte and reach the cathode, where they recombine with oxygen molecules and with the electrons entering the cathode to form water molecules.

As in an electrolytic cell, in an FC the redox reactions also take place at the electrodes: the electrons exit from the cell from the anode through the external load and reenter in the cell at the cathode.

The FC embeds a control system which regulates gases input and output flows, checks and acts upon the reaction temperatures, pressure, and humidity levels, in order to achieve the best operating conditions.

Main advantages of FCs are high efficiency, especially when in CHP mode, low operating pressure and temperature ranges (from 80 to 1000°C, lower than the 2300°C of internal combustion engines), scalability with no decreases in efficiency rates, low environmental impact, performances independent from load variations, high reaction speeds in response to changing loads, and mechanical construction with no moving parts. Main disadvantages are though represented by the necessity to employ scarce and expensive materials, such as platinum as the catalyst, and a relatively short service life of the cells. FCs are categorized according to the types of electrolyte they use and their ranges of operation temperature.

Alkaline Fuel Cell

In an *alkaline fuel cell* (AFC) the electrolyte is generally potassium hydroxide in a water solution with the operating temperature between 60 and 100°C. The electrodes are made of porous carbon and the catalysts are made of nickel and platinum.

At the porous anode, hydrogen spontaneously reacts with OH^- ions in the solution to form water and releases electrons as per the following reaction:

$$H_2 + 2\,OH^- \rightarrow 2\,H_2O + 2\,e^- \tag{2.27}$$

with an oxidation potential of: $\Delta U_{oss} = -\Delta U_{rid} = 0.83$ V. At the cathode, the electrons re-enter the external load and spontaneously react with oxygen and water to produce OH^- ions:

$$\frac{1}{2}O_2 + H_2O + 2\,e^- \rightarrow 2\,OH^- \tag{2.28}$$

with a reduction voltage of $\Delta U_{rid} = -\Delta U_{oss} = 0.40$ V. The OH^- ions inside the electrolytic solution move toward the anode to close the internal circuit. Cell voltage is the sum of $\Delta U_{oss} + \Delta U_{rid}$ of the two reactions, which equals $0.83 + 0.40 = 1.23$ V. Depending on the cell operating temperature, water is released either as liquid or vapor as the only emission from the cell.

The strengths of this type of cell are its fast start-up speed and high efficiency but the cell has a low tolerance for carbon oxides, therefore AFC needs to be supplied by pure oxygen only, which though restricts its range of applications and adoption.

Phosphoric Acid Fuel Cell

Phosphoric acid fuel cells (PAFCs) employ phosphoric acid as the electrolyte, and electrodes made of gold, titanium, and carbon with platinum to avoid corrosion. These cells have a good tolerance for CO_2 and therefore ambient air can be used instead of pure oxygen.

The reaction taking place at the anode is the following:

$$2\,H_2 \rightarrow 4\,H^+ + 4\,e^- \tag{2.29}$$

while the reaction at the cathode is:

$$O_2 + 4\,H^+ + 4\,e^- \rightarrow 2\,H_2O \tag{2.30}$$

Operating at a temperature between 160 and 220°C, these cells can are suitable for use as CHP units.

As PAFCs are CO_2 tolerant, they can also be powered by hydrocarbons and alcohols via external reforming to avoid injection of CO that would damage the platinum catalyst.

Normally, the dimension of a commercial plant using PAFC is between 100 and 200 kW, but plants of several MW have been built in Japan and in the United States to power facilities like hospitals, offices, schools, and airport terminals.

Polymeric Electrolyte Membrane Fuel Cell

Polymeric electrolyte membrane fuel cells (PEMFC, PEM) consist of two porous carbon electrodes separated by a polymeric membrane of chlorinated sulfuric acid. This technology works following the same electrochemical reactions as the PAFC. Ambient air can replace pure hydrogen as the fuel, provided that no CO residues are present to cause permanent damages to the catalyst. PEM operate at temperatures between 80 and 120°C.

PEM cells are characterized by a very high reliability thanks to the noncorrosive polymeric electrolyte, its low operating temperatures, and its tolerance for CO. A very important characteristic is the high current density (around 2 A/cm^2) that grant the cell to be compact and light, have fast start-up speed and reaction to load variations. This is why such cells are very suitable for nonstationary applications, even more because of their simplicity of installation and ease of operations and maintenance. In stationary applications, PEM can generate electricity and function as a power back-up device in systems with a capacity that normally range up up to hundreds of kilowatts. Standardization in construction and the economies of scale from the large number of applications make their cost one of the cheapest within the FC family of technologies.

Molten Carbonate Fuel Cell

The electrolyte in *molten carbonate fuel cells* (MCFCs) is a carbonated alkaline liquid solution (normally KCO_3) in molten state with operating temperature between 600 and 850°C.

The reactions at the anode are:

$$3\,H_2 + 3\,CO \rightarrow 3\,H_2O + 3\,CO_2 + 6\,e^- \tag{2.31}$$

$$CO + CO_2 \rightarrow 2\,CO_2 + 2\,e^- \tag{2.32}$$

while at the cathode the reaction is:

$$2\,O_2 + 4\,CO_2 + 8\,e^- \rightarrow 4\,CO_3^- \tag{2.33}$$

in which the conducting ions are CO_3^-. The reactions at the electrodes correspond to the combustion of H_2 and CO with the production of CO_2 at the anode. Thanks to this feature, the cell can avoid having separate supplies of gases leading to a simplified construction, very high conversion ratios, and suitability to be used in CHP applications. However, the battery has to sustain considerable thermal stresses that potentially reduce reliability and service life, but high operating temperatures can reduce the use of catalysts therefore helping in cutting down manufacturing costs.

At the moment the MCFC seems not suitable for nonstationary uses, but its capability of being powered by different types of fuels makes it extremely flexible and used in a number of commercial stationary applications.

Solid Oxide Fuel Cell

Solid oxide fuel cells (SOFCs) are built with a type of solid and ceramic electrolyte consisting of zirconium oxide (ZrO_2) stabilized with yttrium oxide (Y_2O_3), which caters for very high operating temperatures (between 800 and 1000°C). At such temperature ranges, the electrodes need no catalysts and the SOFC can be powered by a mixture of hydrogen, carbon oxide, and hydrocarbons.

The anodic reactions are:

$$H_2 + O^- \rightarrow H_2O + 2\,e^- \tag{2.34}$$

$$CO + O^- \rightarrow CO_2 + 2\,e^- \tag{2.35}$$

$$CH_4 + 4O^- \rightarrow 2\,H_2O + CO_2 + 8\,e^- \tag{2.36}$$

The reaction at the cathode is:

$$3\,O_2 + 12\,e^- \rightarrow 6\,O_2^- \tag{2.37}$$

in which O^- is the conducting ion.

These reactions correspond to the combustion of hydrogen, oxygen, carbon oxide, and methane; unfortunately, the functioning of the SOFC entails the release of carbon dioxide in the atmosphere [Eq. (2.36)].

This technology entails the adoption of a particular geometry disposition for the cell, which is normally arranged in a tube-like shape. SOFC are usually employed for large stationary applications, since SOFC not suitable for small facilities or nonstationary uses, at least for the time being. The cell start-up time to reach the operating temperature is longer if compared to other types of FCs, but CHP uses are characterized by good performances and high efficiency. As usual, high operating temperatures increase thermal stresses and therefore reduce reliability and service life.

Thermodynamics

While the reaction in an electrolytic cell is not spontaneous, the energy conversion in an FC is spontaneous with a negative Gibbs energy variation. Compared to an electrolytic cell, the reactions are the same but occur in the reverse direction.

In standard conditions, the variation of Gibb's energy is $\Delta G = -237$ kJ/mol. At nonstandard conditions the ΔG is computed as:

$$\Delta G = \Delta H - T\,\Delta S \tag{2.38}$$

where ΔH (the enthalpy change) and ΔS (the entropy change) are obtained by:

$$\Delta H = \Delta H_{H_2O} - \frac{1}{2}\,\Delta H_{O_2} - \Delta H_{H_2} \tag{2.39}$$

and by:

$$\Delta S = \Delta S_{H_2O} - \frac{1}{2}\,\Delta S_{O_2} - \Delta S_{H_2} \tag{2.40}$$

Every cell produces a maximum ideal electric work W:

$$W = \Delta G = q\,E = z\,F\,E \tag{2.41}$$

where z is the number of the electrons involved in the reaction, F is the Faraday's constant, and E is the electromotive force generated by the cell, also called the *reversible voltage* of the cell U_{rev} when it represents the maximum theoretically obtainable voltage in open circuit conditions.

From the previous equation, the reversible voltage of the cell can be calculated by:

$$U_{rev} = \frac{\Delta G}{z\,F} \tag{2.42}$$

Similarly, the total energy supplied by the cell ΔH is linked to the thermoneutral voltage U_{th} through:

$$U_{th} = \frac{\Delta H}{z\,F} \tag{2.43}$$

In standard conditions, U_{rev} equals 1.229 V, while in nonstandard conditions, the reversible voltage is obtained from the value of ΔG calculated in operating conditions.

Temperature and pressure influence the reversible voltage U_{rev} by the following:

$$\left(\frac{dU_{rev}}{dT}\right)_{p\,=\,cost} = \frac{\Delta S}{z\,F} \tag{2.44}$$

$$\left(\frac{dU_{rev}}{dp}\right)_{T\,=\,cost} = -\frac{\Delta V}{z\,F} \tag{2.45}$$

where ΔV represents the change in volume.

When the concentrations of the reactants are different from the stoichiometric value (i.e., excess of hydrogen or oxygen), U_{rev} changes according to the Nernst's equation:

$$U_{rev} = U_{rev}^0 + \frac{R\,T}{z\,F} \log\left(\frac{1}{[H_2]\,\sqrt{[O_2]}}\right) \tag{2.46}$$

in which $[H_2]$ and $[O_2]$ are the concentrations of the hydrogen and oxygen molecules. An increase in pressure augments the reversible voltage by facilitating the transport phenomena, the increase of gas solubility in the electrolyte and the reduction of material loss caused by evaporation. As a trade-off, higher pressures increase mechanical stresses. An increase in operating temperatures reduces the ohmic, activation, and concentration polarizations. The effect of pressure and temperature on cell performance is shown in Fig. 2.2.

The relationship between the electric current and the molar flow rate of the hydrogen consumed by the cell are derived by the Faraday's laws. In stoichiometric conditions, $\dot{n}_{H_2,id}$:

$$\dot{n}_{H_2,id} = \frac{I_{fc}}{z\,F} \tag{2.47}$$

where I_{fc} is the current measured at the electrodes.

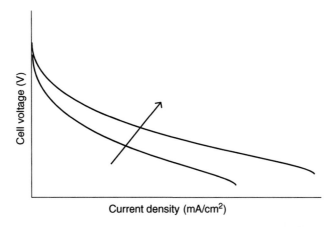

FIGURE 2.2 Example of influence of temperature and pressure on the *I–U* characteristic of the FC.

Because of the losses caused by polarizations, not all the injected hydrogen reacts completely, and the cell consumes a higher quantity of hydrogen than in ideal conditions. In order to quantify this consumption excess after polarizations in real-life conditions, the Faraday's performance can be computed as:

$$\eta_F = \frac{\dot{n}_{H_2,id}}{\dot{n}_{H_2}} \tag{2.48}$$

where $\dot{n}_{H_2}$ is the molar flow rate of the hydrogen necessary to generate the current needed by the load. As a result, the hydrogen consumed in an FC can be calculated as:

$$\dot{n}_{H_2} = \frac{N_c \, I_{fc}}{\eta_F \, z \, F} \tag{2.49}$$

Therefore, the molar flow rates of the oxygen needed for the reaction to occur and the water yielded by the reaction are:

$$\dot{n}_{H_2O} = \frac{1}{2} \dot{n}_{O_2} = \dot{n}_{H_2} \tag{2.50}$$

For a hydrogen/air FC, since oxygen constitutes 21% of the mass of the air, the amount of atmospheric air required for the cell to function is about 4.76 times the quantity of the pure oxygen.

The electric efficiency η_e of an FC is defined as:

$$\eta_e = \frac{U_{fc}}{N_c \, U_{th}} \tag{2.51}$$

where U_{fc} is the output voltage of the cell electrodes.

The maximum current I_{max} obtainable by the FC is given by the Faraday's laws as:

$$I_{max} = n \, F \, \frac{df}{dt} \tag{2.52}$$

where $\frac{df}{dt}$ is the theoretical maximum consumption speed (in mol/s) of the reactants. The actual current flowing in the FC I_{real} is given by an equation similar to Eq. (2.52) where $\frac{df}{dt}$ is computed using the actual velocities of the reacting species.

If the molar hydrogen inflow is $\dot{n}_{\text{H}_2,\text{in}}$ and $\dot{n}_{\text{H}_2,\text{used}}$, the actual molar flow of hydrogen actually reacting (in mol/s), η_e becomes:

$$\eta_e = \frac{I_{\text{real}}}{I_{\text{max}}} = \frac{\dot{n}_{\text{H}_2,\text{used}}}{\dot{n}_{\text{H}_2,\text{in}}} \tag{2.53}$$

For a FC receiving excess hydrogen, the electric current performance is lower than unity. The FC which consumes all the hydrogen inflow (such FC called *dead-end cells*) (having no hydrogen leakage) can reach an efficiency virtually as high as 100%.

Mathematical Model

The polarizations of the FC always increase the voltage at anode and lower that of the cathode, as in:

$$U_{\text{anode}} = U_{\text{rev,anode}} + \left|\eta_{\text{anode}}\right| \tag{2.54}$$

$$U_{\text{cathode}} = U_{\text{rev,cathode}} - \left|\eta_{\text{cathode}}\right| \tag{2.55}$$

in which η_{anode} and η_{cathode} represent the sum of the concentration and activation polarizations related to the anode and the cathode, respectively. The cell voltage is:

$$U_{\text{cell}} = U_{\text{cathode}} - U_{\text{anode}} - R\,I_{fc} = U_{\text{rev}} - \eta_{\text{conc}} - \eta_{\text{act}} - \eta_{\text{ohm}} \tag{2.56}$$

The cell voltage dependency with current density shows the effects of the polarizations as in Fig. 2.3.

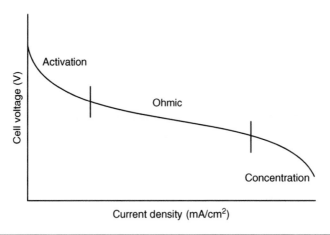

FIGURE 2.3   *I–U* curve and polarization effects.

The I–U curve of the cell can be modeled by measuring the cell output at different operating conditions. An example of equation describing U_{fc} is given by:

$$U_{fc} = U_{fc,0} + C_{1,fc}\, T_{fc} + C_{2,fc}\, \ln\left(\frac{I_{fc}}{I_{fc,0}}\right) + \frac{R_{fc}\, I_{fc}}{T_{fc}} \qquad (2.57)$$

in which $U_{fc,0}$, $C_{1,fc}$, $C_{2,fc}$, $I_{fc,0}$, and R_{fc} are the parameters that are obtained by regressions on experimental data, T_{fc} is the cell temperature, U_{fc} is the stack voltage, and I_{fc} is the current generated in the FC. The first two terms represent the thermoneutral voltage of a series of cells in ideal conditions; the third term embeds the losses caused by activation polarization, while the fourth contains the losses of ohmic polarization. Same as in the model of the electrolyzer, in an FC the losses of concentration polarization are determined empirically.

The operating point of the electrolyzer is determined by solving the nonlinear system between Eq. (2.57) and the electric power provided by the fuel cell $P_{FC} = U_{FC}\, I_{FC}$. The value of I_{fc} obtained from the solution of the system allows the calculation of the hydrogen flow rate $\dot{n}_{H_2}$ used in the FC.

2.2.4 Hydrogen Storage

While using electrolyzers to obtain hydrogen from electricity and FCs to retrieve electricity are a mature and environmentally sound method to manage energy from renewable energy sources, the storage of hydrogen for the short, medium, and long term presents several technical challenges that makes its widespread adoption and use as an energy vector much more difficult [15, 16]. The classic methods to store H_2, like compression or liquefaction, have drawbacks or trade-offs that mine their effectiveness and large-scale deployment. For example, compression storage at reasonable pressures has low volumetric energy density, while high pressure compression result in expensive reservoirs with potentially serious safety risks. Liquefaction, on the other hand, needs so much energy to maintain the pressure and temperature conditions that overall system efficiency is seriously diminished, whereas continuous evaporation of the liquefied hydrogen from the tank limits the use of this technology to applications with high consumption rates. Other new technologies are actively investigated and are showing promising characteristics very suitable for hydrogen storage: nanocarbons and metallic or chemical hydrides potentially result in storage systems with low operation pressures, low safety risks, high charging–discharging speeds, all at very low manufacturing costs.

Compression

Compression storage is considered the simplest technology for hydrogen storage. Due to the low volumetric density of hydrogen though, storage by compression entails building systems capable of achieving high pressures (from 250–300 up to 700 bar) and/or large volumes. Low pressure compression can be applicable to large-scale stationary uses, since the reduced number of stored moles of gas can be compensated by the large volume of ample storage tanks; in nonstationary applications; however, it is necessary to accept a trade-off between high pressure, low volumes, as well as keeping the weight down to a minimum. Compared to methane, hydrogen compression storage also requires a volume three times the volume of methane, at the expense of an amount of specific energy (in MJ/kg) much higher than what needed for methane compression. Hydrogen also needs higher compression pressures due to its lower volumetric energy density.

Commonly used compressors include reciprocating pistons, rotary compressors, centrifugal, and axial turbines. Such compressors must be constructed with compatible materials suitable for the contact with hydrogen. The storage tanks are usually made of aluminum reinforced with fiberglass or polymeric materials with carbon fibers, permitting a gravimetric density around 2–5%.

The typical compressors used are volumetric or dynamic compressors. The ideal work for compression $L_{comp,id}$ of a dynamic, single-stage compressor can be described by a *polytropic compression*[4] as:

$$L_{comp,id} = \frac{m\,R\,T_{in}}{m-1}\left[1-\left(\frac{p_{out}}{p_{in}}\right)^{\frac{m-1}{m}}\right] \tag{2.58}$$

where m is the exponent of the polytropic compression, T_{in} is the temperature of the gas entering the compressor, and p_{in} and p_{out} are the pressures at the input and the output points of the compressor. The temperature of the gas T_{out} at the outlet can be obtained through a mathematical calculation from the expression of the polytropic process and from the law of perfect gas:

$$p_{in}v_{in}{}^{m} = p_{out}v_{out}{}^{m} \rightarrow$$

$$p_{in}\left(\frac{RT_{in}}{p_{in}}\right)^{m} = p_{out}\left(\frac{RT_{out}}{p_{out}}\right)^{m} \rightarrow$$

$$p_{in}{}^{(1-m)}T_{in}{}^{m} = p_{out}{}^{(1-m)}T_{out}{}^{m} \rightarrow \tag{2.59}$$

$$T_{out} = T_{in}\left(\frac{p_{in}}{p_{out}}\right)^{\frac{1-m}{m}}$$

The power used by the compressor P_{comp} is:

$$P_{comp} = \frac{\dot{n}_{gas}\,L_{comp,id}}{\eta_{comp}} \tag{2.60}$$

where $\dot{n}_{gas}$ is the molar flow rate of the gas passing through the compressor and η_{comp} is the efficiency of the compressor.

Given an ideal gas, the pressure p_s inside the storage tank equals:

$$p_s = \frac{n\,R\,T_s}{V} \tag{2.61}$$

where n is the number of moles of the gas in the storage tank, T_s is the tank temperature, and V is the tank volume. The pressure p_s therefore can be calculated from T_s and n, which depends on the molar flow difference between the inlet gas $\dot{n}_{in,s}$ and the outlet gas $\dot{n}_{out,s}$. The temperature T_s is also influenced by the temperature of the gas entering the tank $T_{in,s}$ and by the exiting gas temperature $T_{out,s}$. Therefore, the calculation of p_s requires a system of three nonlinear equations for the three unknowns: p_s, T_s, and $T_{in,s}$.

---

[4] A thermodynamic transformation is defined as *polytropic* when it follows the law $pv^{\gamma} = $ constant, in which γ is the characteristic exponent (or the characteristic number) of the polytropic.

For stationary applications, the most convenient solution is to adopt large-volume tanks to maintain low storage pressures and keep energy consumption down; for non-stationary uses, instead, it is better to employ smaller tanks at the highest possible pressure.

Liquefaction

Liquefaction of hydrogen is a classic and mature technology widely used in the chemical industry since many years, with volumetric density that can reach 50 kg/m^3 and gravimetric density close to 20%. However, the very low temperatures of liquefaction are difficult to maintain, not only because it comes at the expenses of energy which reduces storage efficiency, but also because at the low boiling point of hydrogen, even a very small heat exchange with the environment can cause evaporation inside the tank, requesting to vent gaseous hydrogen to avoid internal overpressurization.

Due to the low critical point of hydrogen, the energy consumed by compression and thermal managements can make the costs of liquefaction much too high. The energy for H_2 liquefaction is around 3.2 kWh/kg, while for N_2 liquefaction is only 0.2 kWh/kg. Around 30% of the energy used (in terms of low heating value) is needed for the liquefaction process, while only 4% is needed for the compression. This represents a problem for small-scale non-stationary applications, particularly in automotive uses, as in order to retain liquefaction of hydrogen, energy is wasted even when the vehicle is parked.

The classical industrial liquefaction process is based on the *Joule–Thompson effect*, that occurs when a gas forced through a valve changes temperature after expansion in adiabatic conditions. A gas is first compressed at ambient temperature then cooled in a heat exchanger; it is then released through a throttle valve where it undergoes the Joule–Thompson effect that causes the liquefaction of a part of the gas; the gas that remains after the partial liquefaction re-enters in the cycle at the heat exchanger.

The industrial-grade *Linde's cycle* is based on the Joule–Thompson effect and is used with different types of gases, such as nitrogen, which cools down at ambient temperature when undergoing an iso-enthalpic expansion. Hydrogen, on the contrary, heats up in an iso-enthalpic expansion. This is why in order for hydrogen to cool upon expansion, its temperature must be below its inversion temperature.[5] For this, a precooling of H_2 to 78 K is performed before hydrogen is sent to the throttle valve. Under the inversion temperature, the internal interactions of the H_2 molecules cause the gas to do work when it is expanded. Some versions of the Linde's process use liquid nitrogen to precool the hydrogen before it passes through the expansion valve; the nitrogen is then retrieved and reused in the process. A possible alternative to the Linde's cycle is the *Claude's cycle*, in which some of the pressurized hydrogen is diverted to an engine to undergo an additional iso-enthalpic expansion.

At standard conditions, hydrogen spin isomers are nearly 25% para-hydrogen and 75% ortho-hydrogen. Since para-hydrogen is in a lower energy state than is ortho-hydrogen, the spontaneous conversion from ortho to para-hydrogen releases heat that introduces hydrogen boil-off losses. Therefore, hydrogen needs to be in its para form for

---

[5] At a given pressure, a nonideal gas has an *inversion temperature* above which the expansion of the gas in an iso-enthalpic transformation causes a temperature increase, while below such inversion temperature an expansion causes a temperature decrease.

long-term storage, rather than in the ortho form. This must be accomplished by pretreatment of hydrogen gas with catalysts that perform such conversion before hydrogen is liquefied.

The reservoirs normally take the form of a sphere to guarantee the lowest surface to volume ratio, while to minimize thermal exchanges of conduction, convection and radiation, the tanks are built with an inner and outer vessel with a gap either kept in vacuum or filled with liquid nitrogen at 77 K.

Physical and Chemical Adsorption

Adsorption is the surface phenomenon that occurs between the *adsorbate* atoms, ions, or molecules that form bonds with the atoms of the surface of the *adsorbent* [17–24]. When the nature of these bonds belongs to Van der Waals' forces (with an adsorption enthalpy of 20 kJ/mol, the same dimension as the enthalpy of condensation), the phenomenon is called *physical adsorption* or *physisorption*. In this case, the adsorbed molecules maintain their nature since no chemical reaction occurs; anyway the adsorbate atoms' three-dimensional shape can be distorted by the forces of attraction exerted by the atoms belonging to the lattice of the adsorbent. If, on the contrary, the interaction at the separation surface of the adsorbent results in a chemical reaction between adsorbate and adsorbent (i.e., a covalent bond results), the adsorption is called *chemisorption*.

Physisorption is a very fast process and does not need high activation energy to form bonds between the adsorbates and the adsorbents. Normally, the adsorbates are in the form of gases and adsorbent are solids. The easiest adsorbed gases are those which are highly polarizable and condensable. In physisorption, the quantity of the adsorbate is proportional to the specific surface of the substratum and the pressure and temperature conditions at which the process occurs. Physisorption also behaves differently according to the surfaces and the materials involved with the possibility of producing multiple layers of adsorbates.

Chemisorption is slower than physisorption and has an enthalpy equal to the enthalpies of the formation of chemical bonds (200–400 kJ/mol). The adsorbed molecules tend to maximize the coordination number[6] with the adsorbent and the length of their bonds is usually shorter than that in a physisorption. Chemically, the adsorbed molecules can be broken and remain separated on the surface of the adsorbent, making the surface a potential catalyst for other chemical reactions.

The opposite phenomenon in which the adsorbed molecules are released from adsorption is called *desorption*.

Empirical models have been developed to describe the interatomic or intermolecular interactions in the phenomenon of adsorption. One of the most commonly employed models is the *Lennard–Jones potential*, describing the energy potential of the interaction between atoms or molecules, considering the Van der Waals' forces occurring at longer distances together with the repulsion forces that happen at shorter distances, and the

---

[6] The *coordination number* means the number of molecules and ions linked to a central atom in a structure. In crystallography, the term is used to indicate the number of atoms directly adjacent to a single atom in a definite crystalline structure.

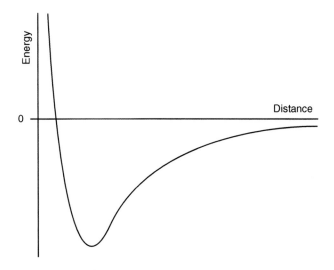

FIGURE 2.4 Lennard–Jones potential for diatomic molecules.

Pauli exclusion principle.[7] The energy potential, or interaction energy, between two atoms, also known as the *12-6 potential*, is expressed as:

$$V(r) = 4\epsilon \left[\left(\frac{\sigma}{r} \right)^{12} - \left(\frac{\sigma}{r} \right)^{6} \right] \tag{2.62}$$

where ϵ is the depth of the potential well, r is the interatomic or intermolecular distance, and σ is the collision diameter of the atoms determined by the kinetic theory of gases. The Van der Waals' forces at long ranges are described by the r^{-6} term, while the repulsion forces at short ranges are expressed by the r^{-12} term. Figure 2.4 shows the trend of the 12-6 potential energy for diatomic molecules (like hydrogen).

The *fractional coverage* θ is defined as the ratio between the number of the sites occupied by the adsorbate and the number of the sites available for adsorption. It can be defined as:

$$\theta = \frac{V}{V_\infty} \tag{2.63}$$

where V is the volume of the adsorbate and V_∞ is the volume of the adsorbate corresponding to the complete adsorption of a layer of adsorbate. The *rate of adsorption* $d\theta/dt$ is the change of the fractional coverage over time.

The variation in the fractional coverage as a function of the pressure at a constant temperature is called *adsorption isotherm*. There are different formulations based on different hypotheses to describe how the quantity of the adsorbate changes according to pressure. One classic formulation is the *Langmuir's isotherm*, which is based on the assumptions

---

[7] The *Pauli exclusion principle* states that no two identical fermions can have the same quantum numbers. A *fermion* is a particle that has half-integer spin and follows the *Fermi–Dirac statistics*. Protons, neutrons, and electrons are examples of fermions.

that adsorption is monolayer and there are no other overlapping layers of adsorbed molecules, all adsorption sites have the same probability of being occupied by the adsorbates, the surface of the adsorbent is perfectly uniform, and the probability of a molecule being adsorbed in a site is independent from whether the adjacent spaces have already been occupied by other molecules.

The hypotheses on which the Langmuir's isotherm is based can cause significant divergences from the experiment results. For example, the assumption on the site equiprobability of adsorption contrasts with the adsorption enthalpy measures that show how such enthalpy declines when the functional coverage increases. For this reason, other isotherms based on less restrictive hypotheses have been introduced.

For example, the *Temkin's isotherm* is expressed as:

$$\theta = c_1 \ln(c_2 \, p) \tag{2.64}$$

with c_1 and c_2 as the constants determined by experiments. This isotherm assumes that the enthalpy varies linearly with the pressure. Another isotherm, assuming that the enthalpy varies logarithmically with pressure is the *Freundlich's isotherm*, given by:

$$\theta = c_1 \, p^{\frac{1}{c_2}} \tag{2.65}$$

If the adsorption occurs with condensation above the first adsorbate layer, the commonly used isotherm in this case is the one developed by Brunauer, Emmett, and Teller (*BET isotherm*):

$$\frac{V}{V_{\text{mon}}} = \frac{c\,z}{(1-z)[1-(1-c)\,z]} \tag{2.66}$$

where

$$c = \exp\left(\frac{\Delta_{\text{des}} H^0 - \Delta_{\text{vap}} H^0}{R\,T}\right) \tag{2.67}$$

V_{mon} is the volume of the adsorbate considering only the first stratum, $z = \frac{p}{p^*}$, and p^* is the vapor pressure of a layer of adsorbate thicker than a molecule. Due to the fact that there can be more than one stratum of adsorbate, the curve does not saturate like a monolayer isotherm, but grows indefinitely instead.

Apart from a few exceptions, the adsorption isotherms can be categorized according to the Brunauer's classification into five types. The isotherms of type I describe the typical monolayer adsorption of the Langmuir's isotherm. Usually chemisorption is a monolayer phenomenon and follows the curve of type I. Figure 2.5 provides an example of Type I physisorption of nitrogen microporous activated carbons.

Type II and III correspond to multilayer adsorptions. The fractional coverage of type II increases rapidly at the beginning, then tends to saturate within a range of pressure values, before rising again with a nearly exponential trend. Type III instead manifests an exponential tendency for all pressure values. Types IV, and V describe adsorptions on porous substrata. Type IV isotherm initially shows a curve similar to type II, but after a certain pressure value it saturates to a limited coverage value. Type V starts with an exponential trend but ends up saturating in a way similar to type IV isotherms.

Some of the materials used for physisorption are, for instance, activated carbons, boron oxide, and zeolytes.

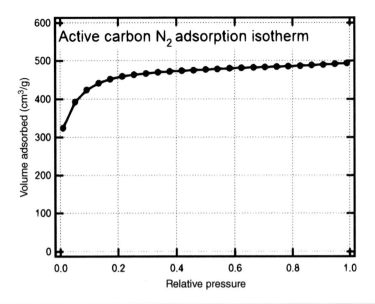

FIGURE 2.5   Type I adsorption profile for N_2 on activated carbons (modified from [25]).

When hydrogen accumulates in the crystalline structure of a metal, a *metal hydride* is formed. In this case, the adsorption of hydrogen is facilitated by decreasing temperature and increasing pressure; hydrogen is released when the opposite occurs, namely heat is increased or pressure reduced. The process therefore is reversible and develops in two phases:

1. *hydrogenation*; hydrogen is injected in the reaction device containing the metallic hydride to diffuse and to become attached to the hydride lattice. This process occurs at 3–6 MPa with the release of heat around 15 MJ/kg (exothermic adsorption).

2. *dehydrogenation*, by increasing the temperature to above 500°C, hydrogen can be restored to its previous biatomic structure and released for reutilization (endothermic desorption or extraction).

One of the most common metal hydrides for hydrogen storage is based on magnesium, where the compound obtained is MgH_2. Since magnesium is an abundant and cheap element, this hydride has a low manufacturing cost, although MgH_2 is characterized by a slow adsorption–desorption kinetics and high operating temperatures (in the range of 300°C), which makes their use for H storage suitable for stationary applications more than for nonstationary applications. Anyway, fast heating technologies like inductive heating, can be employed to speed up cycling, while research is looking for new hydrides capable of operating with fast cycling at lower temperatures (under 100°C). When integrated into an FC system, it is of course possible to retrieve part of the thermal energy produced during the cell operation for the thermal cycles of the hydride storage system.

The hydrogen adsorption (charging) process can be divided into three phases. The first phase (α) is the hydrogen diffusion in the metal structure, given by:

$$\sqrt{p} = K_S\, x \tag{2.68}$$

where K_S is the Sievert constant.

In the second phase ($\alpha + \beta$), hydrogen begins to react with the metal and, regardless of the increase of the concentration of the gas, the pressure remains the same. The final phase (β) occurs when the pressure starts to climb while the concentration of the gas continues to increase. The equations describing the second and the third phases are:

$$\ln(p) = \frac{a}{T} + b \tag{2.69}$$

with:

$$a = \frac{\Delta H^{\alpha \to \beta}}{x\,R} \tag{2.70}$$

where b is a parameter determined by experiments. The process is reversible and can be expressed as the following for a given metal M:

$$M + \frac{x}{2}\,H_2 \leftrightarrow MH_x + \Delta H^{\alpha \to \beta} \tag{2.71}$$

where MH is the metallic hydride, x is the ratio between the quantities of hydrogen atoms and the metal atoms, and $\Delta H^{\alpha \to \beta}$ is the enthalpy of formation of the metallic hydride.

Chemical hydrides can also be used to store hydrogen; in this case, hydrogenation occurs as the result of a chemical reaction. An example of chemical hydride is lithium boro-hydride ($LiBH_4$), solid at standard conditions, which contains 18.5% hydrogen by mass with a reversibility, meaning the percentage of hydrogen that can be retrieved, of 13.8%. Therefore, chemical hydrides can store hydrogen with very good capacities, although not all the hydrogen can be released because part of it remains bound to lithium. The reaction that takes place is:

$$LiBH_4 \leftrightarrow LiH + B + \frac{3}{2}\,H_2 \tag{2.72}$$

where solid LiH cannot decompose further as it is very stable; in fact its decomposition temperature (573 K) is higher than the fusion temperature of $LiBH_4$ (550 K).

Another interesting chemical hydride is sodium borohydride ($NaBH_4$); it is soluble in water, and using ruthenium as catalyst the reaction is the following:

$$NaBH_4 + 2\,H_2O \to 4H_2 + NaBO_2 + 300\ kJ \tag{2.73}$$

with a resulting gravimetric density of 7.5%. Hydrogen can be retrieved by directly de-composing sodium borohydride ($NaBH_4$) as per the following reaction:

$$NaBH_4 + 8\,OH^- \to NaBO_2 + 6\,H_2O + 8\,e^- \tag{2.74}$$

The sodium borate can be further employed to regenerate sodium borohydride.

An important advantage of this type of technology is its long length of storage capacity, which extends for more than 100 days, with operating costs that are comparable to

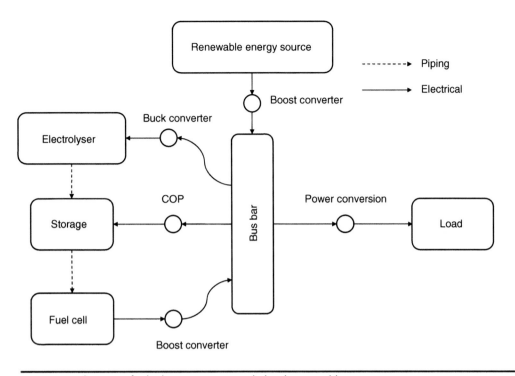

FIGURE 2.6 Schematic of a hydrogen system coupled with renewable energy sources.

traditional fuels. Being liquid, it can be indeed employed in the existing infrastructure for safe and cheap transport and distribution. Sodium borohydride has already applications in the aerospace and automotive industries.

2.2.5 System Architecture and Applications

Renewable energy sources can be used to provide the electrical energy in hydrogen storage plants according to the system schematic shown in Fig. 2.6. Power from REN is connected by a boost converter to the bus bar that works as the backbone in DC mode. By means of a buck converter, the bus bar provides power to the electrolyzer, the output of which enters the hydrogen storage unit. When needed, hydrogen is supplied to the FC by adequate piping. All systems that are needed to condition the storage unit operations are simplified under an intermediate unit characterized by a *coefficient of performance* (COP). Since all these different devices in the system introduce losses according to their respective operating characteristics, the overall efficiency loss of the whole system can become significant; therefore, to achieve the highest possible system efficiency, a control logic must be designed to maximize efficiency by taking into proper considerations the many different operating set-points and technological characteristics of the various subsystems [2].

Many projects have been built and operated around the world, which show that it is easy to combine sources like photovoltaic, biogas, wind, and hydroelectricity together with hydrogen systems. This means that different countries, according to their respective natural context, can opt for the best combination of REN for their hydrogen energy infrastructure. For instance, northern countries with less direct solar radiation will find wind energy or hydroelectricity preferable instead of PV.

Devised as an exposition and information center to inform local communities on sustainable lifestyles and renewable energy technologies, the *GlasHusEtt* is an advanced building pilot project located in Stockholm, Sweden. Alongside traditional heat and electric systems, the building sports a solar hydrogen system that has been designed to use energy from a photovoltaic plant, used in combination with a biogas system [26].

The PV system has a capacity of 3 kW$_p$, using polycrystalline modules and covering a roof area of 25 m^2. The electrolyzer is connected in parallel to the PV plant and the electric grid, being capable of working both in DC and AC modes. The FC can use hydrogen from a reformer or from the tank storage, with a capacity of 50 l. A water tank is used as a heat accumulator to recover the heat from the FC. The system is monitored by a control unit and a monitoring unit which records the system main parameters, like H$_2$ mass flow, electricity, heat, and the meteorological data received from the sensors installed over the building roof.

An example of an industrial renewable energy and hydrogen plant is the one built in Germany by a joint venture between firms of the like of Bayernwerk AG, Siemens AG, BMW AG, and Linde AG [27]. This large-scale plant has been used as a pilot project to gain experience on best practices in the design and operation of hydrogen technology.

The systems that have been designed and integrated in the plant range from photovoltaic generation from different module technologies (like mono-, poly-, and amorphous silicon modules) in fields of of different sizes ranging from 6 to 135 kW$_p$, two low pressure electrolyzers of 111 and 100 kW capable of hydrogen yield of 47 m^3/h, compressors and gas-treatment plants, heaters (supplied by a mixture of gas and hydrogen), and refrigerators (supplied by hydrogen) of 16.6 kW, and three FC, an AFC of 6.5 kW electric and 42.2 kW thermal, a PAFC of 79.3 kW electric and 13.3 kW thermal, and a PEMFC at 10 kW supplied by direct air to power an hydrogen-operated forklift. Hydrogen is also used in a filling station for hydrogen-powered cars.

The knowledge acquired from the plant has shown that safe and efficient operations of utility-scale hydrogen plants systems are possible and current technology is mature and reliable for large-scale commercial deployment.

As for the energy supply fluctuations caused by renewable energy sources, the volatility in power injection can be efficiently smoothed out by the use of hydrogen energy, allowing for a even higher penetration of electricity from REN in the T&D grids. Since hydrogen systems are able to respond quickly and efficiently to the changes in load conditions, hydrogen power systems can perform well as peaking power plants as opposed to other traditional power plants that need hours of ramp-up before starting to inject energy into the grid.

2.3 Methane

Converting and storing renewable electricity in the form of methane and injecting it in the natural gas grid is a viable approach for long term storage. The existing natural gas infrastructure can indeed be used with no needs for difficult and costly pipeline upgrades or the construction of new gas transport infrastructures to accommodate for hydrogen's small molecule transport issues. Methane can be used directly in turbines for electricity generation, for heating, and cooling purposes, or as an alternative fuel option for nonstationary applications. The production of methane from hydrogen and carbon-based species is called *methanation*.

2.3.1   Methanation

One of the most frequently used processes to obtain methane from hydrogen- and carbon-based species is the reversible, exothermic *Sabatier's reaction*, where H_2 combines with carbon dioxide (CO_2) as in:

$$CO_2 + 4\,H_2 \rightarrow CH_4 + H_2O \tag{2.75}$$

with a $\Delta H = -165.0$ kJ/mol. The conversion process takes place in a reactor where a catalyst, like ruthenium or nickel, makes the reaction possible at a temperature range that is between 300 and 400°C and a pressure of 3 MPa.

The Sabatier's reaction takes place by starting with a reverse exothermic water gas shift reaction:

$$CO + H_2O \rightarrow CO_2 + H_2 \tag{2.76}$$

with a $\Delta H = -41.16$ kJ/mol. The carbon dioxide from reaction (2.76) then reacts with hydrogen, completing the reaction (2.75) with the final output of methane and steam.

Carbon dioxide can come from a number of different sources, like ambient air, as a by-product of other industrial processes, or from the sequestration of carbon dioxide as an alternative to venting it in the atmosphere.

The reactors where the process takes place can be either *adiabatic fixed bed reactors* or *isothermal fluidized bed reactors*. Fixed bed reactors run the risk of damaging the catalyst in case there is no sufficiently homogeneous thermal distribution inside the reactor chamber, while ambient conditions inside the fluidized bed reactors can be more difficult to control than fixed bed reactors. Two adiabatic reactors are connected in series with an intermediate gas cooling step to help in achieving better temperature distributions in the two reactors. To help with the correct stoichiometry of the reaction, the first reactor receives an additional flow of CO_2. Due to their simplicity of operations, most of the reactors used for methanation are fixed bed.

An adiabatic fixed bed reactor could be modeled as an ideal *plug flow reactor* (PFR). In an ideal PFR, the chemical reactions are assumed to take place in a cylindrical tube that the reactants traverse in a continuous flow with concentrations that vary along the axis of the tube but are assumed to be perfectly mixed in the cylindrical volume perpendicular to the axis and having infinitesimal thickness. In a PFR, the change in concentration of a chemical species is given by the following equation:

$$C_A(x) = C_{A,0}\,\exp(-k\,\tau) \tag{2.77}$$

where $C_{A,0}$ is the concentration of species A at the inlet of the reactor, k is the rate constant of the reaction (form the law of mass action) taking place inside the reactor, and τ is the residency time according to:

$$d\tau = dx/v_{axial} \tag{2.78}$$

with v_{axial} the velocity of the flow inside the reactor along the axial direction.

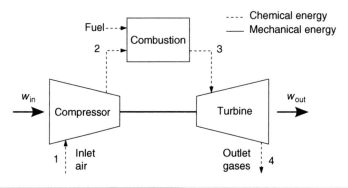

FIGURE 2.7   System schematic for the Brayton cycle.

2.3.2   Combustion in Combined Gas-Vapor Cycles

Electrical energy can be retrieved from the chemical energy embedded in methane (and hydrogen) by means of *Combined Gas-Vapor Cycles* in *combined cycle gas turbine* (CCGT) plants. Such plants combine a gas-fired turbine, to generate electric energy, with a steam turbine, that captures the resulting waste heat from the gas-fired turbine, producing steam that is sent to a downstream steam-turbine, therefore, generating additional electric energy thus increasing the overall system efficiency. These plants are based on the Brayton's and Rankine's cycles, and their very high efficiency makes them one of the best choices to exploit chemical energy storage.

The Brayton Cycle

The *Brayton's cycle* is one of the most commonly employed methods to obtain electrical energy from the combustion of gases [28]. The chemical energy stored in the gaseous fuel (i.e., methane) is converted to thermal energy in a combustion chamber, where gas is mixed with normal ambient air, previously heated up and pressurized by a compressor, and burnt at constant pressure. The gases resulting from the combustion are expanded in a turbine where thermal energy is converted into mechanical energy; an electrical generator, mechanically connected to the turbine shaft, transforms the kinetic energy from the rotation of the shaft into electricity.

Normally, the exhaust gases are vented into atmosphere and are not recovered, making the Brayton's to be classified as an *open* cycle (Fig. 2.7).

The thermodynamical transformations of the system are the following: from points 1 to 2, an isoentropic compression; from 2 to 3, a constant pressure heating; from 3 to 4 an isoentropic expansion; and from 4 back to 1 (equivalent to open air venting to close the cycle), a constant pressure cooling.

The P–v and T–s diagrams for the ideal Brayton's cycle (where all processes are reversible) are shown in Fig. 2.8. The area encompassed within the cycle is the net work $\Delta W = W_{out} - W_{in}$ performed by the system during the Brayton's cycle.

To simplify without a significant loss of information, under the acceptable hypothesis of steady-flow processes that reduces the variations of the gas flows kinetic and potential energy close to zero, the *steady-flow energy balance equation* is given by:

$$(q_{in} - q_{out}) + (w_{in} - w_{out}) = h_{out} - h_{in} \tag{2.79}$$

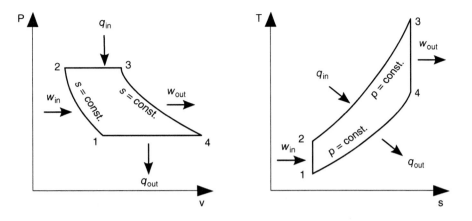

FIGURE 2.8 The P–v and T–s diagrams for the Brayton cycle.

where q is the thermal energy, w is the mechanical energy, and h is the enthalpy, all with the subscripts in and out to indicate if provided to or supplied by the system, and intended per unit mass. Since heat transfer occurs at constant pressure conditions, the thermal energies supplied to and provided by the gas flows are given by:

$$\begin{cases} q_{in} = h_3 - h_2 = c_p\,(T_3 - T_2) \\ q_{out} = h_4 - h_1 = c_p\,(T_4 - T_1) \end{cases} \tag{2.80}$$

where c_p is the specific heat at constant pressure.

As a consequence, the thermal efficiency of the ideal Brayton's cycle is:

$$\eta_{th,\,B} = \frac{w_{net}}{q_{in}} = 1 - \frac{q_{out}}{q_{in}} = 1 - \frac{c_p(T_4 - T_1)}{c_p(T_3 - T_2)} = \frac{1}{r_p^{(k-1)/k}} \tag{2.81}$$

obtained by substitution under the isentropic and constant pressure ideal assumptions, and where $k = c_p/c_v$ is the *specific heat ratio*, $r_p = P_2/P_1$ the *pressure ratio*. k is normally equal to 1.4, since it is the typical value for air at ambient temperature. $\eta_{th,\,B}$ increases if r_p and k increase. Since higher pressures mean higher temperature at the outlet of the combustion chamber, r_p must be limited to avoid damages to components like the turbine blades. In common turbine design, r_p is in the range of 11–16, with $\eta_{th,\,B}$ in the range of 40–60%.

The ratio between the work used in the compressor and the work generated in the turbine is called *back-work ratio*. Normally, in gas turbines such ratio is over 50%,

In a CCGT plant, the Brayton's cycle combines with a Rankine cycle to yield a higher overall system efficiency.

The Rankine Cycle

The *Rankine's Cycle* has been used in many applications in the past (i.e., in rail-road steam engines) as well as today (i.e., in thermoelectric energy plants) [28]. An ideal Rankine cycle can be performed in a system as in Fig. 2.9 and described by the T–s diagram in Fig. 2.10. In the transformation between states 1 and 2, the fluid undergoes an isoentropic compression, followed by an isobaric heating from state 2 to 3. In the transformation

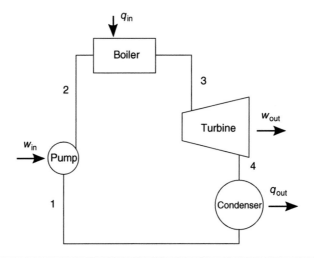

FIGURE 2.9   System schematic for the Rankine cycle.

from state 3 to state 4, the superheated vapor expands isoentropically in a turbine and provides the work that will be converted in electrical energy. Finally, from 4 to 1, the low-pressure wet vapor fluid is condensed in a constant pressure condenser. The heat-transfer fluid can be water or low-boiling organic fluids like halogenated hydrocarbons.

The cooling medium in the condenser can be water, and heat exchangers can take advantage of local rivers or lakes, or air if water is unavailable (*dry cooling*).

The processes can be considered as steady flow, therefore neglecting the variations in kinetic and potential energy and obtaining Eq. (2.79) as per the Brayton's cycle. The

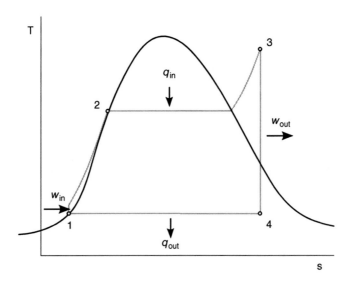

FIGURE 2.10   T–s diagram for the Rankine cycle.

boiler and the condenser do not involve any work, and the pump and the turbine are assumed to be isentropic. Since in the boiler and condenser $w = 0$, and the pump and turbine transformations are isoentropic ($q = 0$), the following relations hold:

$$w_{in} = h_2 - h_1 \simeq v_1(P_2 - P_1) \quad \text{(pump)}$$

$$q_{in} = h_3 - h_2 \quad \text{(boiler)} \tag{2.82}$$

$$w_{out} = h_3 - h_4 \quad \text{(turbine)}$$

$$q_{out} = h_4 - h_1 \quad \text{(condenser)}$$

The thermal efficiency for the Rankine's cycle is given by:

$$\eta_{th,R} = \frac{w_{net}}{q_{in}} = 1 - \frac{q_{out}}{q_{in}} \tag{2.83}$$

where

$$w_{net} = q_{in} - q_{out} = w_{out} - w_{in} \tag{2.84}$$

so that the area encompassed within the cycle is equivalent to the net work performed by the system. The energy efficiency is typically in the range of 30–50%; a way to maximize efficiency is to increase the difference between the temperatures in the boiler and the condenser.

2.3.3   Combustion in Fuel Cells

Another possibility is the use of methane, or syngas mixtures, in FCs like SOFC. High-temperature operations of such FC technology makes possible the conversion of chemical energy contained in methane or syngas (i.e., composed of $CH_4/H_2O/CO_2/O_2$) via a tri-reforming process [29]:

$$CH_4 + H_2O \rightarrow CO + 3\,H_2 \tag{2.85}$$

$$CH_4 + CO_2 \rightarrow 2CO + 2\,H_2 \tag{2.86}$$

$$CH_4 + \frac{1}{2}O_2 \rightarrow CO + 2\,H_2 \tag{2.87}$$

Hydrogen and carbon monoxide are then fed to the SOFC for electricity production and grid injection. The high temperature of the SOFC can also be used in a CHP configuration to increase overall system efficiency.

2.3.4   Applications

Several applications of methanation as energy storage solutions have been studied and simulated, but some have also been constructed as demo plants and are starting to provide valuable pieces of information on its use in the energy infrastructure of the next years.

Wind Energy and Optimized Methane Production

Methanation is a viable energy storage technology for the optimization of the functioning of a large-scale wind farm plant [30]. In this project, the wind farm comprises of 120 turbines each of nominal power of 1500 kW, located in the Gulf of Cadiz, Spain with wind speeds at a height of 80 m in the range of 6–9 m/s depending on the months of the year. Water electrolysis is performed with an AE running at 80° and 100 kPa. Both O_2 and H_2 flows contain water and traces of the other gas; therefore, oxygen undergoes dehydration in a zeolite molecular sieve, and hydrogen is treated in a deoxygenation step with the production of water at a 90°.

CH_4 is obtained from a reverse steam reforming. To avoid the deposition of carbon particles on the catalyst during the reverse SMR reactions, all gas phases and relative ratios must be adjusted with compressors and heat exchangers, while water consumption is reduced by reusing water produced alongside methane. Methane is then fed to the existing gas pipeline network.

Except for the winter months, optimization of methane production yields a stable methane production in the range of 0.55–0.85 kg/s. A plant configuration that produces methane as a function of each month variable wind speed can yield around 25,000 t per year.

Smoothing of Off-Shore Wind Energy

Methanation is employed to minimize off-shore wind farm feed-in unevenness in energy injection in the grid [31, 32]. The wind plant nominal power is 400 MW, the electrolyzer has a capacity of 200 MW at a 70% efficiency, and hydrogen tanks can store up to 10^4 kg.

After conversion and storage of REN as methane, a CCGT (with an efficiency around 60%) is used to provide a flat power injection profile. Sizing of gas storage tanks is fundamental in the overall system design, which is shown to be capable of reaching an overall efficiency of 36%.

Energy for methanation is in the range of 22 kWh/kg including thermal losses in the methanation process (with an efficiency around 60%), and about 2.3 kWh for CO_2 recovery to extract 1 kg of carbon dioxide from air.

Combined Production of Electricity and Syngas

The idea of synthesizing methane from hydrogen and carbon can have an interesting application when carbon is obtained from the sequestration of CO_2, whereby PTG improves the reduction of the environmental footprint of unclean energy sources while at the same time adding energy storage capabilities to the same power plants [33].

In this particular application, hydrogen is obtained from water electrolysis using energy from REN; hydrogen enters a hydrogasification reactor where it mixes and reacts with a fluid mixture of water and a pulverized CO_2-coal mix. The result is a gaseous mix of hydrogen, CO, CO_2, and methane (or *syngas*), which is used either directly in a MCFC power unit or converted to methane prior its injection in the gas pipeline.

Prior to its use, syngas must be purified and is converted to methane via the Sabatier's reaction. Thanks to the scalable architecture of the MCFC, the power block is optimized to always operate at nominal conditions. An additional optimization is performed by the reuse of the CO_2 flow that can be recovered from the anode exhausts of the MCFC.

In this plant, syngas is used both as fuel as well as in the production of methane. The optimization algorithm has therefore an additional parameter, which is how much syngas to use as fuel to get electric energy, or how much to inject in the pipeline.

A feeding ratio R_F is defined as the ratio between the syngas used for electrical energy injection and the total syngas production. An efficiency metric, defined as the coproduction efficiency, measures the ratio between the energy embedded in the electrical energy injected in the electrical grid the one in the methane injected in the gas grid, over the total energy embedded in the fuel consumption of coal and hydrogen. R_F is shown to be always higher when the optimization combines both energy flows instead of only one.

References

1. Jentsch M., Trost T., Sterner M. (2014). Optimal use of power-to-gas energy storage systems in an 85% renewable energy scenario. Energy Procedia 46:254–261.
2. Zini G., Tartarini P. (2012). Solar Hydrogen Energy Systems. Springer: Italy.
3. Agbossou K., Chahine R., Hamelin J., Laurencelle F., Anouar A., St-Arnaud J.-M., Bose, T.K. (2001). Renewable energy systems based on hydrogen for remote applications. Journal of Power Sources 96:168–172.
4. Blarke M. B., Lund H. (2008). The effectiveness of storage and relocation options in renewable energy systems. Renewable Energy 7(33):1499–1507.
5. Cox K. E., Williamson K. D. (1977). Hydrogen: Its Technology and Implications. (1) CRC Press: New York, USA.
6. Ledjeff K. (1990). New hydrogen appliances. In: Hydrogen Energy Progress, VIII, T. N. Veziroğlu and P. K. Takahashi (Eds.). Pergamon Press: New York, Vol. 3, pp. 429–444.
7. Aylward G., Findlay T. (1994). SI Chemical Data (3rd ed.). John Wiley & Sons: New York.
8. Incropera F. P., DeWitt D. P. (1990). Fundamentals of Heat and Mass Transfer (3rd ed.) John Wiley & Sons: New York.
9. Roušar I. (1989). Fundamentals of electrochemical reactors. Electrochemicals Reactors: Their Science and Technology, A. Part and M. I. Ismail (Eds.), Elsevier Science: Amsterdam.
10. Ulleberg Ø. (2003). Modeling of advanced electrolyzers: a system simulation approach. International Journal of Hydrogen Energy (28):21–33.
11. Kothari R., Buddhi D., Sawhney R. L. (2005). Study of the effect of temperature of the electrolytes on the rate of production of hydrogen. International Journal of Hydrogen Energy 30:251–263.
12. Leroy R. L., Stuart A. K. (1978). Unipolar water electrolyzers. A competitive technology. Proceedings of the Second WEHC, Zürich, Switzerland, pp. 359–375.
13. Kordesch K., Simader G. (1996). Fuel Cells and their Applications (3rd ed.). VCH Publisher Inc.: Cambridge.
14. Williams M. C., Strakey J. P., Singhal S. C. (2004). U.S. distributed generation fuel cell program. Journal of Power Sources 131:79–85.
15. Ross D. K. (2006). Hydrogen storage: the major technological barrier to the development of hydrogen fuel cell cars. Vacuum 10(80):1084–1089.
16. Züttel A. (2003). Materials for hydrogen storage. Materials Today, Elsevier: USA, September, pp. 24–33.
17. Atkins P., De Paula J. (2006). Atkins's Physical Chemistry. Oxford University Press: UK.
18. Aranovich G. L., Donohue M. D. (1996). Adsorption of supercritical fluids. Journal of Colloid and Interface Science 180:537–541.
19. Bénard P., Chahine R. (2001). Determination of the adsorption isotherms of hydrogen on activated carbons above the critical temperature of the adsorbate over wide temperature and pressure ranges. Langmuir 17:1950–1955.
20. Bénard P., Chahine R. (2007). Storage of hydrogen by physisorption on carbon and nanostructured materials. Scripta Materialia 56:803–808.

21. Chen X., Zhang Y., Gao X. P., Pan G. L., Jiang X. Y., Qu J.Q., Wu F., et al. (2004). Electrochemical hydrogen storage of carbon nanotubes and carbon nanofibers. International Journal of Hydrogen Energy 29:743–748.

22. Jhi S.-H., Kwon Y.-K., Bradley K., Gabriel J.-C. P. (2004). Hydrogen storage by physisorption: beyond carbon. Solid State Communications 129:769–773.

23. Fukai Y. (1993). The Metal-Hydrogen Systems: Basic Bulk Properties. Springer-Verlag: Berlin.

24. Reilly J. J. (1977). Metal hydrides as hydrogen energy storage media and their application. In: Hydrogen: Its Technology and Implications, K. E. Cox and K. D. Wiliamson (Eds.). CRC Press: Cleveland, Vol. II, pp. 13–48.

25. Author: Nandobike, File:Demac isoth.jpg, Site: en.wikipedia.org, Date: September 6, 2007, Licence: Creative Commons 3.0.

26. Hedström L., Wallmark C., Alvfors P., Rissanen M., Stridh B., Ekman J. (2004). Description and modelling of the solar-hydrogen-biogas-fuel cell system in GlashusEtt. Journal of Power Sources 131:340–350.

27. Szyszka A. (1998). Ten years of solar hydrogen demonstration project at Neunburg vorm Wald, Germany. International Journal of Hydrogen Energy 10(23):849–860.

28. Cengel Y. A., Boles M. A. (2004). Thermodynamics, An Engineering Approach. McGraw-Hill: New York, USA.

29. Er-rbib H., Bouallou C. (2014). Modeling and simulation of CO methanation process for renewable electricity storage, Energy, 75:81–88.

30. Davis W., Martín M. (2014). Optimal year-round operation for methane production from CO_2 and water using wind energy. Energy 69:497–505.

31. Bouyraaman Y., Bendfeld J., Krauter S. (2012). Storage systems to optimize the offshore wind farm feed-in fluctuations. 7th International Renewable Energy Storage Conference and Exhibition (IRES 2012), Berlin, Germany, 12–14 November, pp. 1803–1808.

32. Bouyraaman Y., Bendfeld J., Krauter S. (2013). Optimisation of offshore wind farm feed-in fluctuations via power to gas storage systems. EWEA 2013 Annual Event, Vienna, Austria, 4–7 February.

33. Minutillo M., Perna A. (2014). Renewable energy storage system via coal hydrogasification with co-production of electricity and synthetic natural gas. International Journal of Hydrogen Energy 39:5793–5803.

CHAPTER 3

Electrical Storage

Summary. Electrical energy storage technologies are based on the storage of electric charge by means of either capacitance effects or superconducting electromagnetic inductance. Ultracapacitors and superconducting magnetism are power-intensive technologies with short charging and discharging times, fast response and very high efficiency that make them the technologies of choice when power needs to be recovered or supplied quickly. These characteristics make these technologies suitable for a myriad of applications, from automotive to industrial, and to renewable energy.

3.1 Ultracapacitors

3.1.1 Introduction

Capacitors are devices extensively used in electronic circuits, most notably in applications like filters and control circuits. In their simplest form, capacitors consist of a couple of electrically conductive surfaces separated by a dielectric material: when a voltage is applied on the conductors, an electrostatic field is formed by the positive electric charges accumulating on one plate and the negative electric charges on the other.

The energy that is stored in the capacitor is given by:

$$E = \frac{1}{2} C U^2 \tag{3.1}$$

where U is the voltage applied to the capacitor and C is the *capacitance* defined as:

$$C = \varepsilon \frac{S}{d} = \frac{Q}{U} \tag{3.2}$$

with d the distance between the two surfaces of area S, ε the dielectric constant of the insulator between the two electrodes, and Q is the electric charge. A higher dielectric constant, a larger electrode surface area, a shorter distance between the two electrodes, all contribute to increase the energy stored in the capacitor.

A phenomenon similar to capacitance occurs at the interface of a solid with an electrolyte, where a *Helmholtz's Double Layer* of electric charges is formed: one layer comprises of the ions that are adsorbed to the solid surface, while the second layer is composed by the ions in the bulk liquid which are electrostatically attracted by the ions on the solid interface. Since the ions in the second layer can move freely within the liquid due to thermal motion and electrostatic forces, such layer is referred to as the *diffuse layer*. The width of the double layer structure can be extremely thin (in the range of 10^{-9} m),

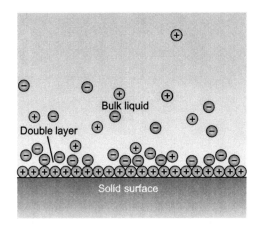

FIGURE 3.1 A Helmoltz double layer forming at a solid–liquid interface [1].

which is much smaller than what could be accomplished in normal capacitors using normal dielectric materials (Fig. 3.1).

The interaction between the two layers results in an electrostatic field similar to the one that is established within a normal capacitor, but the charge density and amount of energy stored in the device depends on the concentration of the electrolyte and on the chemical–physical characteristics of the ions and the solid. In this double layer, the electric field can reach very high values (in the range of 10^5 V/cm). The capacitance, called *differential capacitance* in the *electric double layer* (EDL) formed on the electrode surface is given by:

$$C_\Delta = \frac{d\sigma}{d\psi} \tag{3.3}$$

where σ is the surface charge and ψ is the electric potential on the electrode surface.

An *ultracapacitor* (UC) is a type of electrochemical capacitor that stores the electric energy in the EDL that is established at the interface of two electrodes in contact with an electrolyte. UCs are indeed also referred to as *electrochemical double layer capacitors*. Although UC stores energy by polarization of an electrolyte and therefore could be considered as an electrochemical storage device, since no chemical reactions take place inside the cell, they are hereby considered as an electrical storage technology rather than an electrochemical one.

In a UC, the electrodes are built from porous materials like activated carbons (AC) and coated metal oxides like RuO_2, IrO_2; porous electrodes are indeed employed in order to reach the largest possible interface area between solid and electrolyte. In particular, ACs have an active surface area in the range of 2500 m^2/g and are a chemically inert and corrosion resistant material, therefore improving the life-time stability of the electrodes. The separator is a membrane with high ion permeability used as insulator between the two electrodes, which may be very thin (hundredths of millimeter). Commonly used permeable materials are, for instance, polymeric films. The electrolyte can be organic or water solutions of acids like H_2SO_4. Aqueous electrolytes are mixtures of water as solvent and inorganic acids or alkalies as solutes, with dissociation voltages around 1.15 V per electrode and low internal resistance. Organic electrolytes are organic solvent with salts as solutes. They have a higher dissociation voltage (over 1.35 V per electrode), although can show higher internal resistance. Therefore, since power tends to increase at lower values of internal resistance and the energy density increases with the square of the voltage, aqueous electrolytes can be used in applications where a relatively low energy density and high power density are required; on the other hand, organic electrolytes can be preferable when lower power density and higher energy density are needed. A very important advantage of organic electrolytes lies in the fact that they have lower freezing points than aqueous electrolytes, allowing operations as low as $-40°C$, and as up as $+65°C$.

Since electrodes are made of the same material regardless of them being anodes or cathodes, UC do not have a true polarity as all other kinds of batteries. Nonetheless, UCs are marked with a polarity to make sure that initial polarization, occurring during the first hours of use, is maintained over the long-life cycle of the UC in order to to avoid potential life-time reductions and increased decay that could occur if polarity is reversed.

The large interface surfaces provided by porous materials, combined with the small charge separation of the EDL, provides the buildup in the UC of very high capacitances and storage of electric energy. To increase power and energy volumetric density, the metal foils are rolled up to form a hermetically sealed cylindrical structure, or stacked to form square packaging arrangements.

The two most important parameters that characterize UC are the internal capacitance and resistance. The capacitance of a UC is the sum of the differential capacitances in the two electrodes' interfaces; they can measured by experimental test setups as:

$$C = \frac{I_d \times t_d}{U_n - U_f} \tag{3.4}$$

where I_d is a constant discharge current from the nominal UC voltage U_n to one half of the nominal voltage U_f, and t_d is the time during which such transition occurs.

The equivalent internal resistance of a UC is the series resistances made up by the collector foil (the metal part of the electrode that collect the electrons and connect the UC to the external loads), by the collector foil interface with AC, by the internal resistances of the electrodes, by the ionic losses, and by the separator resistance. Such internal resistance, sum of the series of resistances just discussed, is also referred to as the *equivalent series resistance* (ESR), that is estimated by experiments as:

$$ESR = \frac{U_f - U_{\min}}{I_d} \tag{3.5}$$

where $U_{\min}$ is the UC voltage at the time instant when I_d transitions from its value to zero, and U_f is the voltage reached after a span of time that depends on the time constants of the UC, normally taken as 5 s after the I_d transition to zero. ESR embeds short-term and long-term time constants, and is an estimate of all the resistive behaviors shown in the UC.

3.1.2 Sizing and Modeling

The total energy that can be stored in UC is evaluated by discharging the UC from its nominal voltage to zero. Typically, the energy that is recovered when the UC nominal voltage drops to one half of the nominal voltage amounts to nearly 75% of the UC total energy.

Energy as a function of charge Q is given by [3]:

$$E = \frac{1}{2}\frac{Q^2}{C} = \frac{1}{2}\frac{SOC^2\,Q_{max}^2}{C} = \frac{1}{2}SOC^2\,C\,U_{max}^2 \tag{3.6}$$

where Q_{max} and U_{max} are the maximum charge and voltage and SOC is the state of charge defined by:

$$SOC = \frac{Q}{Q_{max}} \tag{3.7}$$

The total capacitance that is needed for the system sizing is computed by:

$$C_{tot} = \frac{I\,\Delta t}{U_{max} - U_{min}} \tag{3.8}$$

where U_{max} and U_{min} are the maximum and the minimum operating voltage, and Δt is the required cycle time.

To reach the voltage needed for the system design, capacitors must be connected in series as per normal electric theory. The total number of capacitors, assuming that each capacitor as the same nominal voltage U_n, is given by:

$$N_{series} = \frac{U_{max}}{U_n} \tag{3.9}$$

To reach the capacity C_{tot} needed by the system, the number of strings of UC with nominal capacitance C_n to be connected in parallel is:

$$N_{parallel} = \frac{C_{tot}}{N_{series}\,C_n} \tag{3.10}$$

When UC with different capacities are connected in series, *balancing* is needed to avoid that some UC could be charged over their maximum voltage rating. Balancing can be active, with dedicated circuitry, or passive through the use of series resistors.

While there can be many ways to model the behavior of ultracapacitors, the electrical paradigm is one of the most used in the engineering field. Mathematical and non-electrical models, that represent another way to model UC, can yield more accurate results, but run the risk of not being easily generalizable or not being easy to understand and use.

The simplest electrical model to describe UC functioning is a single RC circuit as shown in Fig. 3.2, where R models the ionic and electric components of the total internal resistance, and C the internal capacity of the UC that accounts for the two internal capacitances at the two internal EDLs.

FIGURE 3.2 RC model.

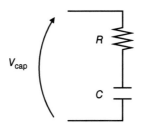

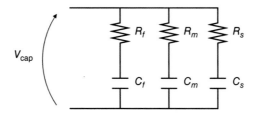

FIGURE 3.3 RC parallel branch model.

In order to consider the nonlinearities of UC voltage, a more complete model can be devised by adding in parallel connections a number of RC branches (a three-branch RC model is shown in Fig. 3.3).

Each branch models UC behavior over different time intervals. The fast-term branch (R_f and C_f) has a small time constant and simulate the UC in the time range under one second, the medium-term branch (R_m and C_m) becomes predominant in the time span of seconds, while the long-term branch (R_s and C_s) comes into action for time ranges of minutes. This refined model attempts to provide a better simulation of the internal ionic transport and exchange mechanisms in the macro, meso, and micro pores of the AC electrodes, therefore, improving the accuracy of the dynamic behavior. These parameters are estimated with relatively simple experimental setups.

Further refinements introduce nonlinearities in the system by adding another branch in parallel to the fast-term capacitor, where the C depends on temperature and voltage of C_f [4]. Another improvement to this model is achieved by an additional parallel resistive branch R_p to take into account the leakage currents, and a series inductance L to cater for high-frequency switching modes (see Fig. 3.4).

Another way to electrically model UC functioning is using the "transmission line model," which assumes that each pore in the AC electrodes can be modeled as a transmission line with its own resistance opposing against the flow of electrons through the AC pores, the EDL is assumed as a set of distributed capacitances, with the final electrical schematic shown in Fig. 3.5.

Thermal heating depends on the duty cycle to which the UC is exposed. The equivalent thermal resistance R_{th} can be determined for each UC device by considering the form factor, the duty cycle, the materials composing its packaging and the eventual heat sinks for convection cooling, or air flow to extract heat from the UC case. A simplified formula for temperature variation as a function of the duty cycle D (a number ranging from

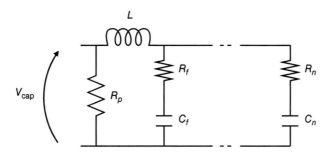

FIGURE 3.4 Generic parallel branch model.

82   Part One

FIGURE 3.5 Three branch transmission line model.

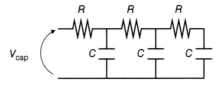

V_{cap}

0 to 1), the current I and the R from Eq. (3.5) is given by [2]:

$$\Delta T = R_{\text{th}} \, D I^2 \, R \tag{3.11}$$

3.1.3   Charging and Discharging

UC are different from traditional electrochemical batteries. Charging and discharging are performed in different ways with different control circuits. UC are characterized by very low internal resistances that, when fully discharged, can be seen as short circuits to most charging sources, by low series inductance, and by small time constants, all of which pose constraints on the type of charging and discharging methodologies.

Normally, two strategies to charge UC are employed: constant current charging, and constant power charging.

Constant current charging is one of the simplest ways to actively control the charge of UC. Normally, boost or buck converters are employed, with the current limit set to the required charge current value and voltage set to maximum UC voltage. Due to the thermal losses of UC as per Eq. (3.11), UC ratings on I^2 limit the maximum current as a function of the duty cycle. In general, fast-switching currents should be avoided since they can cause overheating due to the fast dynamics of the UC.

Constant power charging allows the fastest charging of UC, and is therefore used, for example, in energy recovery from vehicle braking.

A comparison between constant current and constant power charging is given in Fig. 3.6 for a 100 F and 50 V UC module. Constant power charging is nearly twice as fast as constant current, with 145 s lapsed during constant power charging against 250 s in constant current charging.

UC behavior during constant current discharge can be simplified by the following equation:

$$U = U_n - \frac{I}{C} \, dt - I \, R \tag{3.12}$$

where U_n is the nominal UC voltage. An example of a constant current discharge for a 48 V UC for automotive applications is given in Fig. 3.7 for different discharge current values [2].

A way to calculate the discharge time of a UC is to assume capacitance C and resistance R to be constants. This way, discharging is given by [6]:

$$U_0 - U = I \, R + \int \frac{1}{C} \, dq \tag{3.13}$$

where U_0 is the initial voltage at full charge, and:

$$dq = I \, dt \tag{3.14}$$

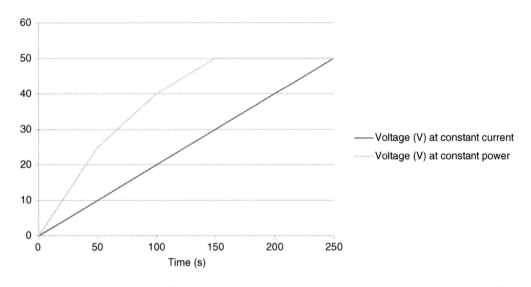

FIGURE 3.6 Comparison between constant current and constant power charging (modified from [5]).

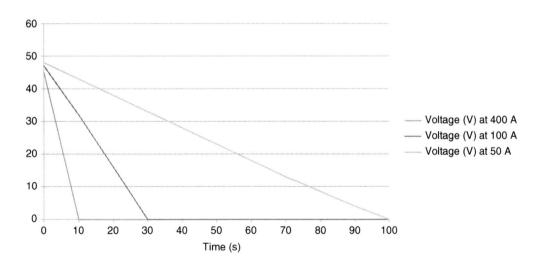

FIGURE 3.7 Discharge of a 48 V UC with different constant currents (modified from [2]).

and

$$I = \frac{P}{U} \tag{3.15}$$

with P the power, and by substitution and after some calculation steps, the time of discharge t_{dis} can be calculated with the following equation:

$$t_{dis} = \frac{1}{2} \frac{C U_0^2}{P} \left[\left(1 - \frac{P R}{U_0^2} \right)^2 - \left(\frac{U}{U_0} \right)^2 \right] + R C \ln \left[\frac{V}{U_0} \frac{1}{1 - \frac{P R}{U_0^2}} \right] \tag{3.16}$$

3.1.4 Efficiency

At the end of the discharge, a UC does not release all the energy that was previously stored: there will be indeed a difference ΔE between the energy that is stored in the UC during the charging process, and the energy released during discharge.

The resulting energy efficiency is given by:

$$\eta_e = \frac{E_{dis}}{E_{ch}} \tag{3.17}$$

where E_{ch} and E_{dis} are computed by:

$$E_{ch} = \int_0^{\Delta t_{ch}} U(t) \, I(t) dt \tag{3.18}$$

$$E_{dis} = \int_0^{\Delta t_{dis}} U(t) \, I(t) dt \tag{3.19}$$

Efficiency is in the range of 85–95%, but efficiency shows a dependency on the operating voltages, decreasing in the range of 86–88% when the gap between the operating voltages is high.

Using the simple RC electric model for the UC to analyze the constant current charging, the voltage across the UC is given by:

$$U(t) = \frac{I}{C} t + U_0 \tag{3.20}$$

where U_0 is the UC voltage at the beginning of the charge. Charging time becomes:

$$t_{ch} = \frac{C \, (U(t) - U_0)}{I} \tag{3.21}$$

The energy lost by Joule effects on the UC internal resistor is given by:

$$E_R = R I^2 t \tag{3.22}$$

while the energy stored is:

$$E_{UC} = \frac{1}{2} C \, (U^2(t) - U_0^2) \tag{3.23}$$

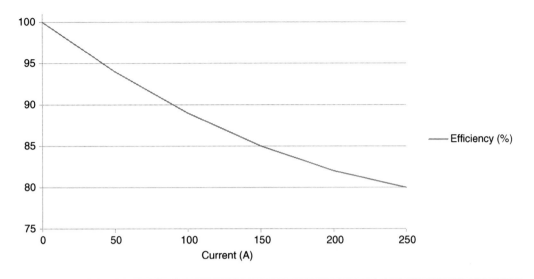

FIGURE 3.8 Constant current charging efficiency.

Since the energy required by the whole process is:

$$E_{tot} = \int_0^t I\,U(t)\,\mathrm{d}t = E_R + E_{UC} \tag{3.24}$$

the charging efficiency for the constant current charging (Fig. 3.8) is:

$$\eta_{ch} = \frac{E_{UC}}{E_{tot}} = \frac{E_{UC}}{E_{UC} + E_R} = \frac{U_0 + I\,t/2C}{R\,I + U_0 + I\,t/2C} \tag{3.25}$$

In constant voltage charging (Fig. 3.9), if $U_0 = 0$, the charging efficiency is given by:

$$\eta_{ch} = 1 - \exp\left(-\frac{t}{RC}\right) \tag{3.26}$$

Typical time constants for UC are in the range of 1 s. Normally, 100% of charge is reached after five time constants with one time constant corresponding, roughly, to a 65% charge and a 35% discharge state.

Tests have shown that over a specific charging current threshold (i.e., 160 A), the constant current charging method has a lower efficiency than voltage current charging, while for lower charging currents, the constant current charging method yields a better efficiency [7].

Performance over time is influenced by the temperature and voltage levels sustained by the UC. Since capacitance and internal resistance change according to how high are the average voltages and temperatures applied over time on the UC. Higher average temperatures will result in higher capacitance decays, that can become as high as 30% over 6000 h of operations at the highest temperature and voltage limits; if the same highest voltage limit is maintained but at ambient temperature, the same capacitance decay in the range of 30% is reached after more than 80,000 h. Similar effects are found on the internal resistance, with increases in R of more than 130% at highest temperature and voltage limits.

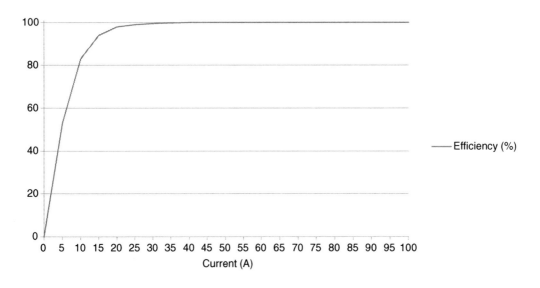

FIGURE 3.9 Constant voltage charging efficiency.

3.1.5 Hybrid Ultracapacitor Systems

UCs are frequently used in coordination with additional energy storage technologies in an hybrid energy storage system configuration. The concept is to use UC as the power-intensive technology providing energy for short periods and quick reaction times, leaving to the energy-intensive technology (usually electrochemical batteries) the task to provide energy for longer periods with longer reaction times. Since the UC are likely the first storage technologies to be requested to assist the load, the control logic always prioritizes charging of the UC over the other storage systems.

UC can be connected in a number of different ways. The simplest and least expensive connection is the *passive hybrid* (see Fig. 3.10, where B, C, and L represent the battery, capacitor and load), where UC and batteries are connected in parallel, and current flows without any control capabilities and low system efficiency.

To improve control over current and power flows, a more sophisticated connection design strategy is the *semiactive hybrid*, which takes advantage of a DC–DC converter connected between the load and the storage system. Three main configurations are usually implemented.

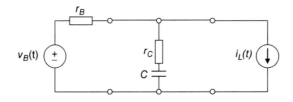

FIGURE 3.10 Passive hybrid (modified from [8]).

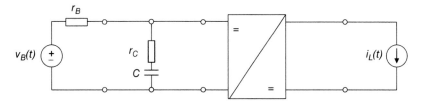

FIGURE 3.11 Parallel semiactive hybrid (modified from [8]).

The first is the *parallel semiactive hybrid* where the DC–DC converter is connected in parallel between the energy storage and the load (Fig. 3.11). Since the UC voltage is still linked to the battery voltage, system performance and control are still suboptimal.

The second is the *capacitor semiactive hybrid* connects the DC–DC converter between the capacitor and the load (Fig. 3.12), improving UC performance by decoupling the load voltage from the UC voltage and making possible a finer control on the UC current.

The *battery semiactive hybrid*, where the DC–DC converter is placed between the battery and the load (Fig. 3.13), granting a constant current discharge of the battery that improves its lifetime and performance, while at the same time avoiding voltage matching between load and battery.

A further improvement in UC hybridized systems is the *active hybrid* (Fig. 3.14) configuration. By using two DC–DC converters, it is possible to completely decouple the battery and the UC from the load for a better overall energy storage performance. The battery converter is designed according to the average load demand, while the UC converter is sized on the peak current the UC is required to provide [8]. This improvement can be sized to have the highest efficiency of all other possible configurations.

The choice between the different connection configurations is to be decided on a case-by-case basis depending on the hybrid system application.

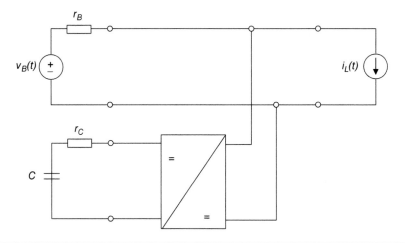

FIGURE 3.12 Capacitor semiactive hybrid (modified from [8]).

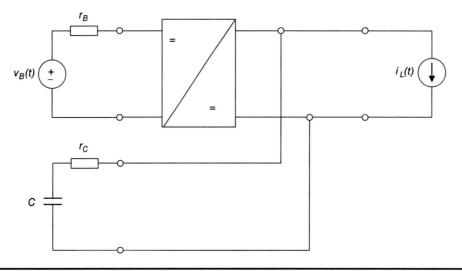

FIGURE 3.13 Battery semiactive hybrid (modified from [8]).

3.1.6 Applications

The combination of virtually unlimited charging–discharging cycles with the wide ranges of specific power (10–10^6 W/kg) and specific energy (0.05–10 Wh/kg) makes these devices a useful technology for many energy storage applications. Advantages and disadvantages of UCs can be singled out as the following [3]:

- life cycle very high, over 10^5 cycles at 75% depth of discharge
- minimal maintenance needs
- loss of capacitance over life cycle less than 20% at average usage conditions

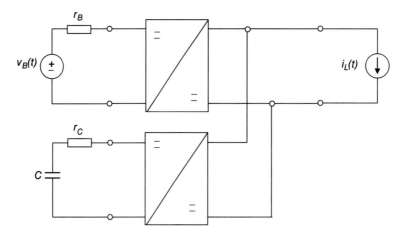

FIGURE 3.14 Active hybrid (modified from [8]).

- efficiency can reach values above 95%, due to the absence of electrochemical reactions and very low internal resistance

- very high power density

- charge and discharge rates are very high

- energy capacity and energy densities are low if compared with other technologies

- self-discharge phenomena can be more significant than in other storage technologies

Another important advantage of UC is that it is very easy to compute its state of charge by logging current and voltage change and discharge over time. The same is not at all easy for many other types of battery technologies.

UC can be employed when there is the need to provide continuity and quality of service to main power supply. In the events when there are short voltage drops and outages in distribution lines, UC can very quickly and efficiently provide power backup to the load. Due to fast charge and discharge rates, all applications that need pulsed currents can be ideally suited for UC. Since no chemical reactions take place inside the UC, they can deliver pulse currents in a very efficient way in all applications requesting fast and short lasting discharge of electric energy to the load. The applications that stem from the peculiarities of UC are therefore varied within different industries. From automotive and transportation to industrial and consumer, UC can be used for the cranking of motors, power steering, energy recovery from regenerative braking, train tilting, UPS, fast smoothing of uneven REN power injection, wind turbine pitching, power backup, telecommunications, forklifts, security doors, memory backups in portable devices, toys, and many other applications [2, 9–12].

In the REN industry, UC are used to supply power for the pitch control system of wind turbines to make sure that the rotor speed remains within its nominal operating specifications. The long life and usability in a wide operating temperature range make them a reliable and relatively low cost option to wind turbines control. Another application is to smooth the output voltage of a wind farm or a photovoltaic plant to avoid excessive voltage sags and spikes due to varying meteorological conditions: UC are charged when the voltage is too high and release energy when a drop in power output occurs. All these events last typically for less than 5 min (and can be even under the 1 min threshold), but the large number of charge and discharge cycles imposed by the unevenness of the RES are well coped with by the UC due to their characteristics.

Industrial applications (i.e., data centers, hospitals, and telecommunications) entail the use of UC banks integrated in UPS to assist the standard battery pack; the UC manages the power-intensive functions while the battery supplies the energy over longer period of times.

In the automotive industry, UC are used as the *kinetic energy recovery system* (KERS), which stores energy during braking and releases it to boost acceleration and reduce fuel consumption. Additionally, KERS reduces the load of the mechanical brakes increasing their life time and reducing their maintenance. Another typical automotive usage is the start-stop application, where UC provide the initial starting current needed to restart the motor after a stop; the large number of start-stops to be faced by the storage system and the very high and impulsive start currents would make the use of traditional batteries impractical.

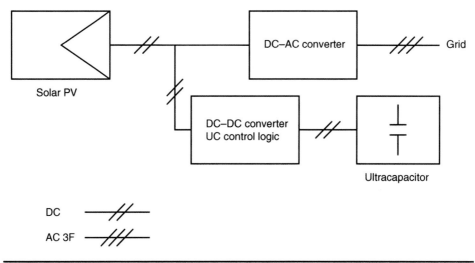

FIGURE 3.15 Example of parallel configuration with battery pack in a PV REN application.

Smoothing of Photovoltaic Energy Injection

Smoothing the uneven power injection of photovoltaic plants is a very interesting applications in the REN industry. A configuration that is used to provide power in the short periods of time when solar radiation is unstable is shown in Fig. 3.15, where PVGS is the PV plant, PIS is the power injection system, comprised of the variable speed inverter (VSI) that converts DC current into AC current and injects energy in the grid, and the DC–DC converter that bidirectionally manages energy flows in and out the UC module [11]. In this configuration, no energy-intensive storage technologies are employed.

Automotive Application with Li-Ion Batteries

An example of an active parallel application with Li-ion batteries for an automotive application is given in Fig. 3.16. The DC–DC converter is controlled via an energy-management supervisor unit; the buck or boost converter connects the UC pack to the motor, in parallel with a Li-ion battery pack that provides the long-term energy to the motor. The UC pack is provided as a three branch transmission line model as discussed earlier in this section. The energy management logic also manages the correct functioning of the storage system by considering all parameters linked to performance, fuel efficiency, state of charge of the batteries, and their optimal functioning [12].

Pulsed Currents

The UC capability to supply high power and high current in short periods of time can be employed to supply loads that require consumption profiles with pulsed constant current characteristics. Many portable electronic devices or electric vehicles have similar load profile requirements. Combining UC with batteries in a hybrid configuration improves system efficiencies and can solve the usual power-to-energy compromise problem that make the use of standard battery technologies not optimal. In these applications indeed, either high power and high energy content are required to the energy storage pack.

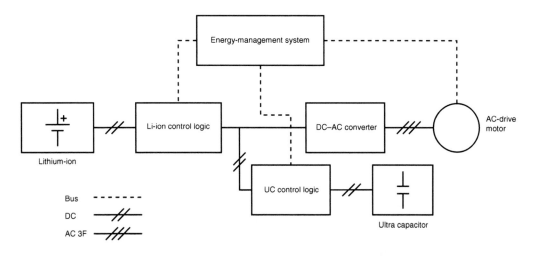

FIGURE 3.16 Example of active parallel configuration of UC with Li-ion battery pack in an automotive application.

Pulsed currents can be decomposed, for ease of analysis, in a constant component $i_{L,\text{ave}}$, and a dynamic component $i_{L,\text{dyn}}$ with zero average value (Fig. 3.17), with a duty cycle D and a period T.

The equation that describes the pulsed current is given by:

$$i_L(t) = i_{L,\text{ave}}(t) + i_{L,\text{dyn}}(t) \tag{3.27}$$

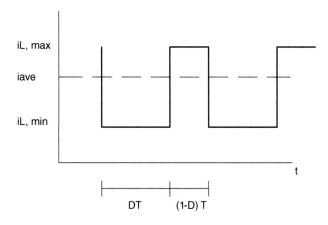

FIGURE 3.17 Typical pulsed current cycle.

where:

$$i_{L,\text{ave}}(t) = \frac{1}{T} \int_0^T i_L(t)\,dt = D i_{L,\text{MAX}} + (1 - D) i_{L,\text{MIN}} \tag{3.28}$$

In the operating cycle, the charge is given by:

$$Q_L = \int_0^T i_L(t)\,dt = Q_{L,\text{ave}} + Q_{L,\text{dyn}} \tag{3.29}$$

where:

$$Q_{L,\text{ave}} = \int_0^T i_{L,\text{ave}}(t)\,dt = (D i_{L,\text{MAX}} + (1 - D) i_{L,\text{MIN}}) \times T \tag{3.30}$$

and

$$Q_{L,\text{dyn}} = \int_0^T i_{L,\text{dyn}}(t)\,dt = 0 \tag{3.31}$$

The maximum dynamic load charge would therefore be:

$$Q_{L,\text{dyn}}^{\text{MAX}} = \int_0^{DT} i_{L,\text{dyn}}(t)\,dt = \int_{DT}^T i_{L,\text{dyn}}(t)\,dt = D T(1 - D) \times (i_{L,\text{MAX}} - i_{L,\text{MIN}}) \tag{3.32}$$

The UC will therefore need to be sized to supply or recover an energy equal to $v_L \times Q_{L,\text{dyn}}^{\text{MAX}}$, where v_L is the voltage applied to the load.

3.2   Superconducting Magnetic Energy Storage

3.2.1   Introduction

Superconducting magnetic energy storage (SMES) technology employs a superconducting coil at cryogenic temperatures to store energy by electro and magnetic induction.

The energy stored in an induction coil is given by:

$$E = \frac{1}{2} L I^2 \tag{3.33}$$

and power by:

$$P = \frac{dE}{dt} = U I \tag{3.34}$$

where L, I, and U are the coil inductance, direct current, and the direct voltage.

While a simple inductor is not able to store energy for a sufficiently long period of time due to the internal resistance and resulting losses by Joule effects, if a superconductive inductor is used instead, its internal resistance is so low that the current can flow with virtually no losses, storing energy in the superconducting coil-magnet assembly. Along as having to be manufactured with special materials, the coil has to be maintained at temperatures below a critical temperature T_c, the current has to be lower than a critical current I_c, and the magnetic field has to be lower than a critical field H_c [13].

The energy stored in a SMES with a circular cross-section coil-magnet assembly is given by [14]:

$$E_0 = \frac{B_0^2}{2\mu_0} \, \pi \, R^2 \, h \, f(2\,R/h) \tag{3.35}$$

where B_0 is the magnetic far field, R is the mean coil radius, h is the axial extension, and $f(2R/h)$ is a factor which corrects the energy by taking into account the near field effect.[1] For other geometries, the correction factors are provided by specialized datasheets. For instance, in toroidal assemblies, h is the mean circumference $h = 2\pi R_0$ with R_0 as the outside radius, and $f = 1$.

Since there are no moving parts, virtually no internal resistance, no Joule losses, and the resulting efficiency in the coil-magnet assembly is very high and close to 100%. The overall system efficiency though must take into consideration the losses in the power conversion stage and the energy needed to maintain the right temperature conditions for the superconductive device to work properly. Final system efficiency therefore, though still high, goes down to 90–95%. In SMES, power density is close to 10^3 W/kg.

3.2.2 System Construction and Control

The main subsystems of an SMES are the superconducting coil, the refrigeration unit and the power conditioning system.

The superconducting coil can be manufactured from low-temperature superconductors (LTS) or high-temperature superconductors (HTS). LTS are materials that need cooling temperatures of liquid hydrogen (20 K) or liquid helium (4 K), while HTS operate at the temperature of liquid nitrogen (77 K). Refrigeration costs between the two technologies are very different, but lower costs for HTS trade off against lower energy efficiency and storage densities. HTS materials are mostly made from ceramics, which are brittle and hard to shape, while LTS coils employ metallic alloys like niobium-titanium.

Normally, coils are designed in solenoidal or toroidal shapes. Solenoids are easier to manufacture but their magnetic field tend to be strong in the far field of the coil, meaning that some caution and distance must be kept when deciding placement of the SMES. Toroidal coils instead manage to contain most of their magnetic field in the near region. Design choice depends on the size of the SMES; for instance, smaller SMES can be made solenoidal, while as the size and power increase, toroidal design becomes preferable to reduce magnetic far field effects.

Refrigeration is provided by a cryostat, a double cooling stage system composed by three layers: the external layer is at room temperature, the middle layer is at the liquid nitrogen temperature, and the internal layer is at the liquid helium temperature. Between the external and the middle layer, there is a vacuum cavity which to reduce convection

---

[1] The *near field* and *far field* are the regions of the electromagnetic field in the vicinity or distance of an electromagnetic body, like a magnet or antenna.

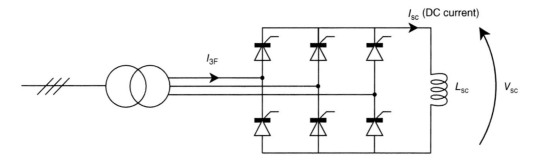

FIGURE 3.18 Thyristor SMES system.

heat exchanges, and a radiation shield against radiation heat transfer. Refrigeration is provided by a *cryo-cooler* using Stirling, Gifford-McMahon, or Joule–Thomson cycles.

The power conditioning system is the unit that manages the bidirectional flow of electrical energy to and from the SMES. There are three main configurations for the power conditioning system which controls the power transfer between the coil and the grid [15].

The simplest design is the *thyristor-based* SMES (Fig. 3.18). It uses a wye-delta transformer, an AC/DC thyristor bridge and a superconductive inductor. The thyristor bridge acts in rectifier mode (charging) when the phase angle α (namely, the angle from the natural switch-on time to when the switch-on command is given to each thyristor) is under 90°, and in inversion mode (discharging) when α is greater than 90°. This configuration, though simple, causes distortion on the output current with high ripple and harmonic content that affect the correct operations of the SMES coil.

The voltage across the coil is:

$$U = U_0 \cos \alpha \tag{3.36}$$

where U_0 is the no-load DC voltage of the bridge. The current and the voltage of the coil are given by:

$$I = \frac{1}{L} \int_{t_0}^{t} U \, d\tau + I_0 \tag{3.37}$$

where I_0 is the initial current of the inductor. Since I is not reversible, the power depends only on V: if it is positive, the power is transferred to the SMES, if it is negative, the power is supplied by the SMES to the load.

The *voltage source converter* (VSC) SMES is composed by a wye-delta transformer, a six-pulse pulse width modulation (PWM) rectifier/inverter using an insulated gate bipolar transistor (IGBT), a two-quadrant DC–DC chopper with IGBT and a superconductive inductor. The PWM and the chopper are connected by a DC link capacitor (Fig. 3.19). The charge and the discharge is controlled by the chopper which control coil voltage by switching on or off the IGBTs. Harmonic content in the current is lower than in the thyristor-based configuration.

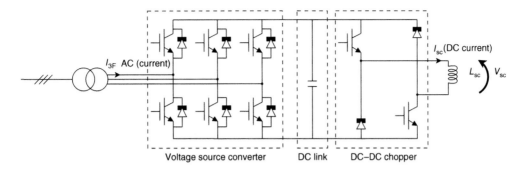

FIGURE 3.19   Voltage source converter SMES system.

Finally, the *current source converter* (CSC) SMES design is shown in Fig. 3.20, where a set of *LC* filters smooth the harmonics from the AC line. This configuration can inject, if needed, reactive power to the grid and ensures low harmonic contents and current ripple to the superconductive coil to reduce losses in the system.

3.2.3   Applications

Thanks to the advantages peculiar to SMES, long life time with virtually no limits to the number of charge–discharge cycles, fast response times, low maintenance from the lack of moving parts, SMES can be applied to a multitude of energy storage applications if feasible for a technology that has high costs from cryogenic cooling and construction with

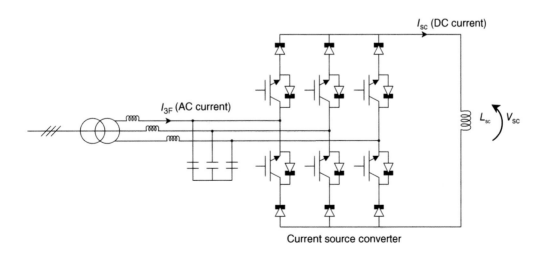

FIGURE 3.20   Current source converter SMES system.

special superconductive materials. Most of actual applications are anyway still demo or test installations, since SMES is a technology still in development or precommercial stage.

SMES can be used to help in grid frequency regulation by combining them into the power conversion system of REN energy plants [16]. The resulting conversion system is equivalent to an ideal synchronous generator that can supply balanced AC three phases with controllable amplitude and phase with the additional advantages of nearly immediate response, no additional impedance, and management of active and reactive power. An application to stabilize the output frequency in a microgrid has been shown using a 640 kJ dual magnet SMES, with power provided either by the grid or by two REN systems, a 20 kW PV system and a 30 kW wind energy turbine, and backup from a 50 kW diesel generator. The grid power can be switched off and the SMES is shown to be capable of sustaining the grid frequency and smoothing power unevenness from REN [17].

A wind energy plant of nominal power 10.2 MW is connected to an AC bus together with an SMES system in a VSC configuration [18]. Fluctuations in the frequency output vary depending on relative power output between wind generation and SMES power capacity. For instance, a 55% SMES power capacity is capable of reducing fluctuations down to 0.006 Hz for a 10% wind power injection, and to 0.012 Hz at a 20% of wind. Lower SMES capacities are not sufficient to smooth frequency instabilities.

A preliminary sizing for a commercial SMES 1 GW, 5 GWh bulk energy storage system give a coil diameter as large as 400 m and an height of 40 m [19]. Sizing has taken into consideration seismic construction requirements for the site of installation in an underground cavern, with a 6 MW cryostat system and an overall efficiency of 90%.

Systems of this size can effectively provide benefits to transmission lines, although costs can represent a disadvantage if compared with similar long-term storage systems like pumped hydro and compressed air energy systems.

SMES can be hybridized to combine with other energy storage technologies for smart grid applications in transmission, distribution and end-user power lines (see Fig. 3.21). For example:

- SMES_A: installed after a generation system to stabilize power output
- SMES_B: in an interconnected network to smooth low-frequency oscillation from different centralized generators
- SMES_C: connected to a transmission line as a flexible AC transmission system for time-of-use applications
- SMES_D: in a distribution line as a distributed flexible AC transmission system for power quality applications
- SMES_E: installed in a DC-link bus to stabilize power output
- SMES_F: connected at end-of-user for power quality applications

Other applications for SMES have been studied and tested. Even energy-intensive applications have been considered as a feasible; load following, black start, spinning reserve.

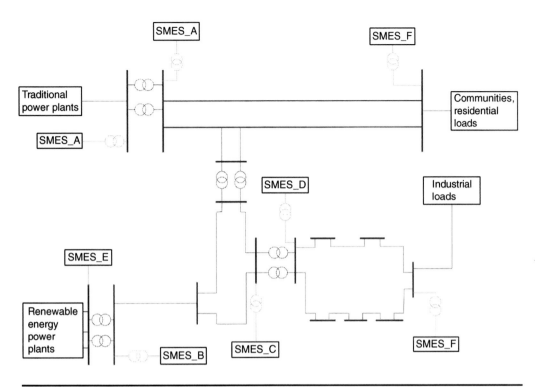

FIGURE 3.21 Applications of SMES in the smart grid.

References

1. http://en.wikipedia.org/wiki/File:Double_Layer.svg, under the Creative Commons Attribution-Share Alike 3.0 license.

2. Maxwell Technologies Inc. (2009). Product Guide, Maxwell Technologies BOOSTCAP Ultracapacitors. Doc. No. 1014627.1.

3. Pavković D., Hoić M., Deur J., Petrić J. (2014). Energy storage systems sizing study for a high-altitude wind energy application. Energy, 76:91–103, 1 November 2014, 10.1016/j.energy.2014.04.001.

4. Shi L., Crow M. L. (2008). Comparison of Ultracapacitor Electric Circuit Models. Power and Energy Society General Meeting-Conversion and Delivery of Electrical Energy in the 21st Century IEEE, Pittsburgh, PA, 20–24 July 2008, pp. 1–6.

5. Maxwell Technologies Inc. (2005). Charging of Ultracapacitors. Application Note 1008981 Rev. 1.

6. Burke A., Miller M. (2011). The power capability of ultracapacitors and lithium batteries for electric and hybrid vehicle applications. Journal of Power Sources 196:514–522.

7. Jiang X., Zhang J., Jian W. (2013). The analysis of ultracapacitor charging efficiency. 2013 International Conference on Computational and Information Sciences, IEEE, Shiyang, pp. 1198–1201.

8. Kuperman A., Aharon I. (2011). Battery-ultracapacitor hybrids for pulsed current loads: A review. Renewable and Sustainable Energy Reviews 15:981–992.

9. Do Y. J., Young H. K., Sun W. K., Suck-Hyun L. (2003). Development of ultracapacitor modules for 42–V automotive electrical systems. Journal of Power Sources 114:366–373.

10. Glavin M. E., Hurley W. G. (2012). Optimisation of a photovoltaic battery ultracapacitor hybrid energy storage system. Solar Energy 86:3009–3020.

11. Minambres-Marcos V., Guerrero-Martinez M.A., Romero-Cadaval E., Milanes-Montero M.I. (2013). Active power injection control of a photovoltaic system through ultracapacitor storage. Industrial Electronics Society, IECON 2013, 39th Annual Conference of the IEEE, Vienna, pp. 5928–5933.

12. Miller J., Prummer M., Schneuwly A. (2009). Power Electronic Interface for an Ultracapacitor as the Power Buffer in a Hybrid Electric Energy Storage System. White Paper Maxwell Technologies.

13. Nielsen K. E. (2010). Superconducting magnetic energy storage in power systems with renewable energy sources. Norwegian University of Science and Technology, Department of Electric Power Engineering, Marta Molinas, ELKRAFT, June 2010.

14. Ries G., Neumueller H. W. (2001). Comparison of energy storage in flywheels and SMES. Physica C 357–360:1306–1310.

15. Hasan Ali M., Wu B., Dougal R. A. (2010). An Overview of SMES Applications in Power and Energy Systems. IEEE Transaction on Sustainable Energy, 1(1):38–45.

16. Molina M. G., Mercado P. E., Watanabe E. H. (2007). Static synchronous compensator with superconducting magnetic energy storage for high power utility applications. Energy Conversion Management 48(8):2316–2331.

17. A-Rong K., Gyeong-Hun K., Serim H., Minwon P., In-Keun Y., Hak-Man K. (2013). SMES application for frequency control during islanded microgrid operation. Physica C 484:282–286.

18. Takahashi, R., Murata, T., Tamura, J., Kubo, M., Kuwayama, A., Matsumoto, T. (2007). Smoothing control of wind power generator output by superconducting magnetic energy storage system. Proceedings of the International Conference on Electrical Machines and Systems 2007 (ICEMS 2007), October 08–11, pp. 302–307.

19. Masuda M., Shintomi T. (1987). The conceptual design of a utility-scale SMES. IEEE Transactions on Magnetics, 23(2):549–552.

CHAPTER 4

Flywheel Mechanical Storage

Summary. Green electric energy can be successfully stored as mechanical energy. One of the oldest ways to store energy is the flywheel, a rotating horizontal wheel that stores kinetic energy in its rotational motion. Electricity can be used to power an electric motor that transfer the motion to a spinning rotor that accelerates inside the flywheel assembly; when the rotor decreases its rotational speed, the system converts kinetic energy to electric energy in the motor now operating in generation mode. Hydroelectric plants, compressed air plants, gravel plants are also capable of storing electric energy as mechanical energy. These technologies will not be described in this chapter, since their use is difficult to be scaled down, and are therefore employed for large-scale installations; nonstationary, or stationary applications for residential, commercial, or industrial power can indeed be limited.

4.1 Introduction

A flywheel is a device with the form of a disk or cylinder, the rotor, capable of storing a significant amount of rotational energy. The ability of spinning objects to store kinetic energy has been known by mankind since the Neolithic age. Many are the applications over the centuries, from the realization of clay vases, to the use of flywheels in early agricultural tractors, to steam trains. When shafted to the electric motor/generator, the flywheel stores energy when the motor increases its angular speed, or releases energy to the generator when the latter acts as a brake to decrease the flywheel angular speed and provides electricity to the load. A *flywheel energy storage system* (FESS) is typically used whenever power is needed over a very short period of time, since FESS do not store large amounts of energy. This is why flywheels are mostly intended and designed for use as power-intensive applications, like in uninterruptible power supply (UPS) systems [1–5].

4.2 Construction and Main Characteristics

The assembly of a flywheel is shown in Fig. 4.1. The motor/generator is enclosed in the flywheel assembly, is directly connected to the flywheel shaft, and to reduce friction losses, the shaft rotates around a set of magnetic or superconductive bearings in a

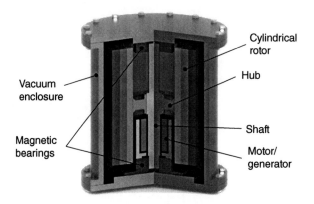

FIGURE 4.1   Example of flywheel rotor assembly [6].

vacuum chamber, so that the system can maintain fast rotation regimes and high power-to-volume ratios. As flywheels are kept in a vacuum during operation, it is difficult to transfer heat out of the system, so a cooling system is usually integrated with the FESS device. To avoid dangerous failure modes in case of breakage of the rotor, an additional containment structure surrounds the FESS to protect from debris projection if a failure occurs.

One of the main advantages of FESS technology is its very fast response time, either in charging or discharging, while low energy density constitutes one of the main drawbacks of their uses. Another advantage of FESS is its capacity to sustain practically unlimited charging and discharging cycles without degradation, and are highly reliable and durable with a service life expectancy of many years. The use of inert materials also renders flywheels more environmental-friendly than traditional batteries. Flywheels though tend to be heavy weight devices with gyroscopic effects that can make their nonstationary uses difficult.

The energy E of a spinning flywheel is given by:

$$E = \frac{1}{2} I \omega^2 \tag{4.1}$$

where ω is the angular velocity and I is the inertia momentum given by the simplified relation for a disk geometry:

$$I = \frac{1}{2} m \left(r_{ext}^2 - r_{int}^2 \right) \tag{4.2}$$

where m is the mass, r_{ext} is the external radius and r_{int} is the internal radius of the cylinder. Increasing the angular speed increases stored energy more effectively than increasing mass. Since increasing ω causes higher friction and thermal losses, the rotor could start facing deformation problems until damages start occurring in its mechanical structure. The tensile strength of the material therefore determines the maximum operational velocity, where the tensile strength σ is defined as:

$$\sigma = \rho \omega^2 r^2 \tag{4.3}$$

and ρ is the density of the material. Lighter materials are characterized by smaller inertial loads at a given speed, therefore composite materials, with low density and high tensile strength, are the materials of choice for storing kinetic energy. In the composite material, the fibers are oriented along the circumference of the rotor; in case of a failure happening during rotation, the cracks would tend to develop along the circumference and correct fiber orientation design becomes important to reduce or avoid fragment projection in case of a catastrophic failure. Always for safety reasons, flywheels have an admissible rotational speed limited to a maximum of 20% of the nominal maximum speed.

The specific energy E_S (kinetic energy per unit mass) is given by the flywheel's geometry and materials, as in:

$$E_S = k \frac{\omega}{\rho} \tag{4.4}$$

where k is the shape factor of the rotor geometry. As an important added advantage, the FESS state of charge can easily be measured since it is given by the rotational velocity of the rotor.

The usable energy density in case at a speed drop s is given by:

$$E_S = (1 - s^2) k \frac{\omega}{\rho} \tag{4.5}$$

In case of $s = 0.2$, the depth of discharge of the flywheel becomes 96%.

Flywheels can be designed for low speed or high-speed operations. A low speed flywheel has advantages of lower cost and the use of proven technologies when compared to a high-speed flywheel system. The first operates at rotations per minute (rpm) in the range of thousands of rpm, while high-speed flywheels rotate with speeds in the range of tens of thousands of rpm. Low-speed flywheels employ steel as the main structural material in the rotor; high-speed flywheels instead use advanced composite material, such as carbon-fiber or graphite.

Auto-discharge is due to friction phenomena that cannot completely be avoided. In FESS, these losses are called *idling losses*. As a result, flywheels need to receive periodical power injection to maintain speed, although these idling losses are usually less than 2% of overall stored energy. A full discharge cycle is possible at any discharge current with typical discharge efficiency of an FESS system up to 97%.

A typical 120 kVA flywheel with 25–30 kWh energy storage has a cylinder-shaped structure with a diameter around 1 m and an height of 2 m.

A list of some of the main parameters of a flywheel are shown in Table 4.1. Apart from the response time speed, most notable is the specific energy as low as 3.3 Wh/kg; this, combined with very fast dynamics, make flywheel use very efficient for power-intensive applications.

4.3 System Architecture and Operations

The flywheel energy storage system has three modes of operation: charging, stand-by (or idle), and discharging mode. The use of fast-switching power electronics makes it possible to operate FESS at high power, fast response times, using standardized industrial control logic network.

Specific energy (Wh/kg)	3.3–5
Nominal power (kVA)	120
Nominal energy (kWh)	0.75–12
Mass (kg)	150–200
Volume (l)	70–100
Voltage (V)	550–850
DOD (%)	80–100
RT efficiency (%)	85
Typical cycle number	> 75,000
Self-discharge (% monthly)	2
Response time (ms)	> 15–20
Full ramp up (ms)	100–500

TABLE 4.1 Typical Flywheel Main Data

The motor/generator of the FESS is typically a *permanent magnet synchronous machine* (PMSM) in a brushless configuration. The conversion and regulation is performed by the power electronics comprising of induction filters, power converters with intermediate DC link (see Sec. 6.5). The control logic regulated the charge and discharge by controlling the speed of the rotor. The transformer connects the FESS to the grid for proper voltage step-up for charging and discharging to and from the distribution grid.

Management of the electromagnetic bearings is one important function of the control logic of a FESS; gap sensors are used to measure the relative position of the rotor shaft and the electromagnets, and a sophisticated control logic ensures that levitation of the flywheel rotor is always stable with no contact ever made between the shaft and the bearings.

The electronics ensure four-quadrant system capability. Overloading of a few seconds is also possible without disruptions or damages to the system; depending on state of charge, reaching 120–150% of nominal power is possible, for time frames in the range of 10 s, either for active or reactive power injection.

In a flywheel energy storage system, charge and discharge are symmetrical, so discharge profiles are the same as charge profiles. Figure 4.2 shows the flywheel AC output during discharge as a function of time; the dashed line represents the locus of points that limits discharge duration as a function of power demand, and the continuous line is the actual power injection profile at a level of power and for the duration limited by the continuous curve. The response times of a flywheel are typically in the range of 20 ms or even less, while full ramp-up from power request to power output is in the range of 500 ms (Fig. 4.3).

4.4   Examples of Applications in Wind and Solar PV Power Plants

Flywheels can be successfully used to smooth power injection in grid-connected PV power plants. A simplified single-line schematic of FESS with PV is shown in Fig. 4.4.

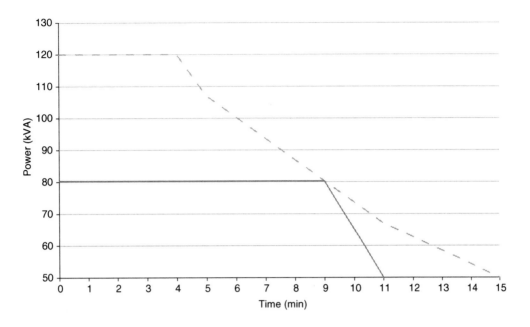

FIGURE 4.2 Flywheel AC power output as a function of time in discharge mode.

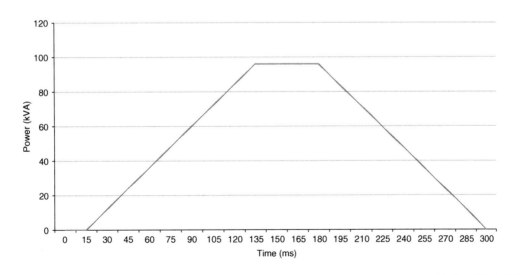

FIGURE 4.3 Example of 120 kVA flywheel power up and down ramping.

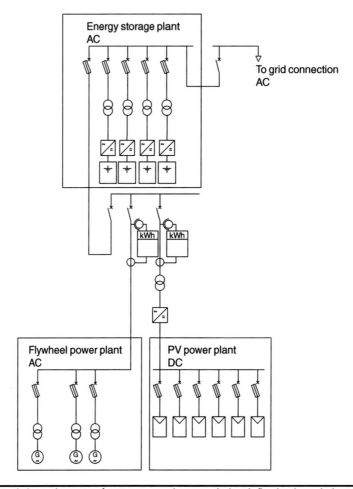

FIGURE 4.4   Single-line schematic of a PV power plant coupled with flywheels and electrochemical energy storage.

The plant control logic monitors the power conversion from the PV plant inverter on the AC side. By using an algorithm based on first and second derivative computations, if the PV power injection goes down, the flywheel is put in discharge mode, and in a time frame of less than 1 s starts injecting its energy in the transformer connected to the grid. This way, the FESS smooths out the reduction in PV power injection and avoids instability on the distribution grid. In discharge mode, the FESS reduces the sags in power injection. When PV power conversion starts rising again, the flywheel continues to inject energy until a threshold is reached and the FESS is put in charge mode. In charge mode, the FESS reduces the upward spikes in power injection. An example of a flywheel for the smoothing of the PV power injection for a real plant operating in southern Europe is shown in Fig. 4.5.

Flywheels can be used with similar results to smooth the inherent energy injection variability of wind energy plants. The use of optimized energy management in a

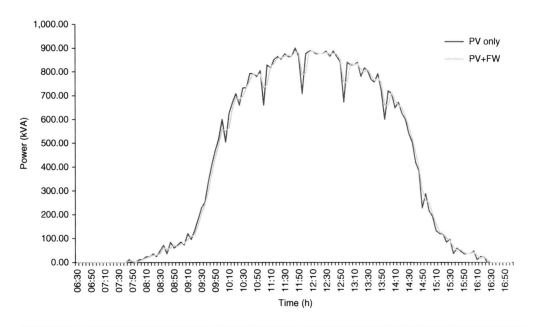

FIGURE 4.5 Smoothing of PV power grid injection for a plant operating in southern Europe.

combined flywheel and wind turbine setup can provide reductions in the range of 90% of the turbulent energy component. For example, a wind turbine of 1.5 MW can be assisted with very good results by a flywheel of 100 kVA [2].

References

1. Diaz-Gonzalez, F., Sumper, A., Gomis-Bellmunt, O., Villafafila-Robles, R. (2012). Modeling and validation of a flywheel energy storage lab-setup. (ISGT Europe), 2012 3rd IEEE PES International Conference and Exhibition on Innovative Smart Grid Technologies, Berlin, pp. 1–6.
2. Díaz-González F., Sumper A., Gomis-Bellmunt O., Bianch F. D. (2013). Energy management of flywheel-based energy storage device for wind power smoothing. Applied Energy 110:207–219.
3. Bjolund B., Bernoff H., Leijon M. (2007). Flywheel energy and power storage systems. Renewable and Sustainable Energy Reviews 11:235–258.
4. Su W., Jin T., Wang S. (2010). Modeling and simulation of short-term energy storage: flywheel. 2010 International Conference on Advances in Energy Engineering (ICAEE 2010), IEEE, Beijing, pp. 9–12.
5. Koshizuka N., Ishikawa F., Nasu H., Murakami M., Matsunaga K., Saito S., Saito O., et al. (2003). Progress of superconducting bearing technologies for flywheel energy storage systems. Physica C (386):444–450.
6. http://commons.wikimedia.org/wiki/File%3AExample_of_cylindrical_flywheel_rotor_assembly.png. By Pjrensburg CC BY-SA 3.0 (http://creativecommons.org/licenses/by-sa/3.0), via Wikimedia Commons.

Energy Storage System Design and Functioning

Summary. All the electricity storage technologies described in the previous chapters are only one part of the overall system that manages the energy flows, from conversion to storage, from power conditioning to energy management and interfacing with the load or the grid. The integration of the energy storage units with the other sub-systems, basics of hardware and software design for the optimization of its operations using the information available on energy electricity prices, load needs, and the integration with energy from renewable plants are outlined in this chapter.

5.1 Initial Considerations

Many energy storage projects tend to be one-of-a-kind endeavors; whether it be for the many different applications that energy storage technology is capable of pursuing, or the many different customers and market conditions, the system integration and the design of the control logic is rarely the same. In any case, all projects should be initiated with an organized mindset that helps in starting a first assessment of project's opportunities and risks.

An initial evaluation should consider the applications that the storage system will be capable of assisting and the related revenues and costs. The economics of the project will not only be a function of the price of the system, the costs of shipment, insurance, site preparation, building construction, installation, electrical connection to the grid or the load, but also the lead time that will take from start of project to final commissioning. In case of financing needs, its arrangement and costs will have to be factored in both in the project time-line and the changes that the project can cause on the firm's structure and cost of capital. Also, it will be necessary to know in details which kind of legal permits or authorizations will the project need to obtain before actual construction.

A preliminary sizing should envisage, for instance, what will be the energy that will be managed by the system, what will be the electricity price historical data and, if available, it forecast to ensure future economic viability of the project, what technology would be best suited for the applications that are targeted by the organization, what will be the technology merits, performances, key process indicators, and reliability of the chosen systems. A prominent piece of information: how many charge and discharge cycles will the technology be capable of sustaining before revamping or complete substitution.

A full contract set for construction, operations, and maintenance (after-sale technical service), with a clear and detailed set of warranties on the performance of the system will have to be negotiated and signed with the supplier of the storage system. For ordinary and extraordinary maintenance, a set of spare parts will have to be kept in stock or anyway readily available for prompt intervention.

A very important thing to consider is the possibility to modify the software of the control logic: how easy or feasible it is to modify it according to changing specifications or customer needs. Having such ease of programming changes can make the difference if the system wants to be robust against changes in law and regulations, or if the system is to be reused in a totally different setting and configuration.

At the end of project life, recyclability of the complete system must be considered to comply with local regulations and to minimize the impact on the environment. Who will be responsible for recycling and at what costs?

After having taken heed of all the points outlined before, and all info collected, the following necessary preliminary technical and financial analysis can begin.

5.2 Energy Storage System Architecture

An ESS is a combination of subsystems that operate together to manage and control the flow of energy to and from the storage unit to fulfill the applications intended to service the connected load or grid (Fig. 5.1). Many ancillary systems are needed for the storage device to work properly. A complete ESS comprises of the following subsystems:

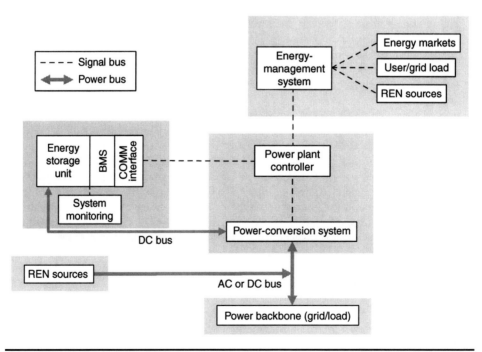

FIGURE 5.1 Schematic of an ESS.

- *Energy storage unit* (ESU). The system that embeds the storage technology (i.e., the electrochemical cell).

- *Battery-management system* (BMS), the control system of the ESU. A communication interface can consolidate the functioning of many ESU so that the storage subsystem can be seen as a single unit by the other ancillary systems. A monitoring subsystem takes care of collecting data from all sensors that are checking the functionality of the ESU.

- *Power-conversion system* (PCS). The power electronic system that converts the power from the ESU to the form of power needed by the backbone where the energy is injected. It can be an AC/DC converter (also called *inverter*) or a DC/DC converter. Usually, inverters are needed for electrochemical or electric storage devices, while are not normally needed for storage systems that directly convert electricity to AC. In this case, the manufacturers can refer to it directly as the *power unit* (PU) or the *power block* (PB).

- *Power plant controller* (PPC). The system that controls the interaction between the ESU and the PCS power conversion logic. The PPC embeds the peculiarities and specificity of the energy storage technology in order to correctly drive the charging and discharging cycles of the ESU.

- *Energy-management system* (EMS). The logic that drives ESS overall functioning. Such logic combines information from energy markets (i.e., prices of electricity during peak and off-peak hours), load needs, info from renewable energy sources productivity, and optimizes the functioning of the ESS according to the specific applications to which the ESS is applied.

- *Electric equipment.* Cables, switches, overvoltage and overcurrent protections, surge protection devices to protect against lightnings, transformers, and current and voltage sensors, protection interface devices to safely connect the ESS to the grid and/or the load.

- *Physical containment.* Mechanical structures, civil works, designed to meet the environmental conditions of a specific site or to make the ESS safe for interaction with operators.

The sub-systems that control and manage the energy stored in the ESU are often referred to as the *balance of system* (BOS). The BOS affects the performance efficiency of the whole system so much that its correct design and good functioning is of the utmost importance. The BOS is fundamental also to expand to many different operation environments and applications that often span over a large range of environmental conditions and specificity of use.

A simplified electric single-line schematic is shown in Fig. 5.2.[1] Two renewable power plants (in the figure, the "PV Power Plant" and the "Wind Power Plant") provide clean energy to the ESS (the "Energy Storage Plant"); an ancillary thermal energy back-up power plant (i.e., a biomass gensets) can be foreseen to provide continuity even in conditions of low irradiation and wind speeds.

---

[1] The meaning of the symbols should be clear to the reader and will not be explained in this text.

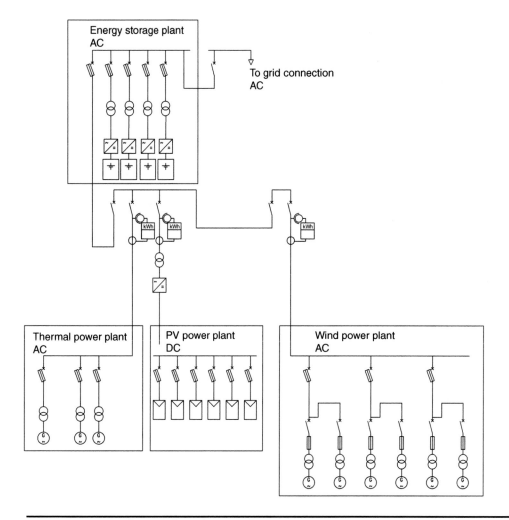

FIGURE 5.2 Example of a single-line schematic of an energy storage system coupled with renewable energy and thermal plants.

The ESS can be connected in the following ways:

- to the DC side of the plant
- to the AC side after the metering unit of the generation unit of the power plant
- to the AC side before the metering unit of the generation unit of the power plant

Figures 5.3 and 5.4 show the possible points of connection of the ESS to a load, the distribution grid, and a generator (e.g., a solar PV or wind turbine plant) when available. Metering units are bidirectional to take into account the flows of energy from the grid to the ESS and backward. In Figs. 5.3 and 5.4, metering units 1 and 3 measure the energy

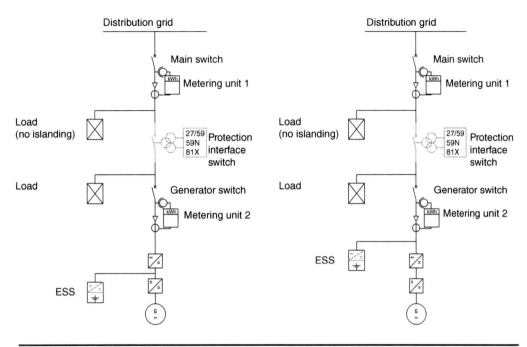

FIGURE 5.3 Possible ESS connections within the power plant.

that is exchanged with the grid, and metering unit 2 the energy that is produced by the generation unit.

5.3 Energy Management System

The software and hardware system that drives the optimized functioning of the overall energy storage system is often referred to as the EMS. Actual implementation of the EMS depends on many factors, but a simplified hardware architecture of an EMS can be identified as in Fig. 5.5.

A router manages the communication of data packets between the ESS and external devices over the internet. For instance, data from the electricity energy price providers, the industrial load, the power output from REN sources, can be exchanged by algorithms like a file transfer protocol over open or virtual private network connections. A network switch connects the systems that are inside the computer network of the ESS, like the inverters, the electric storage units, and the servers. The energy manager software is running on the servers, whereas data that are used by the EMS are managed and stored in databases and data storage units.

The UPS backups the power to make sure that short-term power interruptions are not disrupting the operations of the system and that data are properly saved before system crash. UPS guarantees proper operation even in case of unacceptably bad power quality. Length of intervention is usually in the range of twenty minutes, but coexistence with

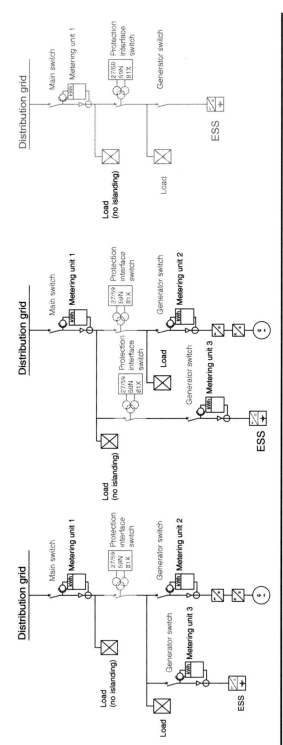

FIGURE 5.4 Possible ESS connections within the power plant.

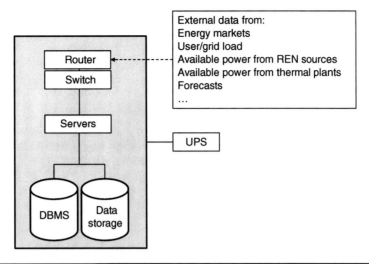

FIGURE 5.5   Simplified architecture of the EMS.

energy storage systems within the same installation can avoid the use of UPS altogether and provide longer back-up periods.

Several different control logic routines are running within the EMS software framework. The energy manager software uses all high-level data, like electricity prices, power requested by the load, availability of power from energy sources, energy storage system status, and finds the optimal set-points to achieve the best possible behavior of the overall system according to the needs of the system user and the applications made available from the ESU.

When interfacing with renewable energy sources that are inherently uneven, forecasts are needed to program and optimize the set-points for charging and discharging according to load needs, which too might be provided as forecast for the next hours or days. Forecasts for renewable energy plants need to consider seasonality, time of day, and of course environmental parameters like temperature, solar irradiation, or wind speed. The system needs to provide active and reactive power to the load or the grid, so forecasting can become a very important software module. Usually, forecasts are needed 1 day in advance up to intervals shorter than 2 or 3 h.

Depending on the dimensions of the system and the economical value of the project, servers can be replicated to cater for redundancy and system stability. If one server fails, a second one takes over with zero or minimum downtime. Together with hardware redundancy, also software redundancy can be provided by running virtual clustering software with the aim to provide immediate recovery in case of software failure.

Lastly, a GUI interface like a SCADA (Supervisory Control And Data Acquisition) software is needed to monitor and control the functioning of the whole system. The SCADA provides communication between the EMS and the other subsystem software, the management of alarms, warnings, events in general, the manual programming or overwrite of signals and registers, the visualization of the current system status, the retrieval of historical data for the benefit of the plant managers and operators.

The EMS does not provide the low-level control of the PCS and the ESS modules; normally, dedicated software is provided by the manufacturers of those systems, where the EMS acts as the provider of the set-points that must be followed by the combination of PCS and ESU as in sections.

5.4   Battery-Management System

The *battery-management system* is the set of subsystems that perform the hardware level management of the storage units, or cells, constituting the ESU (Fig. 5.1).

A list of common BMS functions are the following:

- Regulation
- Overvoltage
- Undervoltage
- Ground-fault
- Short-circuit
- Overcurrent
- Undercurrent
- Temperature
- Pressure
- Balancing
- Data-logging

The BMS monitors, computes, and records the value of parameters that depend on the cell technology, but most commonly are the SOC, the SOH, the temperature and pressures, the maximum or minimum current or voltage limits, the energy and power that have been charged or discharged, the number of cycles from start, and the total amount of energy and power from start of operations. One of the fundamental values that must be checked is the residual charge in each cell, measured in Ah and referred to as the *Coulomb counter*; an example of Coulomb counter for large-scale and small-scale batteries (a cell phone battery) is provided in Table 5.1.

	Large scale	Small scale	Unit
Capacity (Coulomb counter)	8000	3	Ah
Voltage	230	3.8	V
Current	1500	0.375	A
Hours	5	8	h
Power	1035	1.425	W
Energy	5520	11.4	Wh

TABLE 5.1   Example of Coulomb Counters and Main Electrical Parameters

One of the most important functions of the BMS is making sure that the cells function inside the right operating range. This means that BMS must check and regulate electrical parameters like cell voltage or cell current to avoid overcurrent, overvoltage or undervoltage, ground faults and short circuits, leakage current faults, and environmental values like temperature and pressure, that can cause damages or suboptimal functioning of cells.

5.4.1   Cell Balancing and Precharging

Cell imbalance is due to small differences in cell construction that are inevitable during the manufacturing process. Differences can result in different cell behavior during charge and discharge cycles, and can materialize as differences in the open circuit voltages, the self-discharge rates, the internal resistance, the residual charge. Not only the different behavior is caused by manufacturing, but also different working conditions, like heat-exchange differences between cells due to the positioning inside the battery pack, lead also to differences in the performance of each cell. *Cell balancing* (also called *cell equalization*) is the technique that is used in BMS to normalize the cell performance by individually managing the charge and discharge of each cell. Cell balancing increases the total available ESU capacity and extends the cycle life of the ESS.

If the charge or discharge algorithm only checks the series voltage of the overall battery pack to decide on the available SOC of each cell and perform start-stop operations, the ESU will incur in suboptimal functioning. During discharge, the cell that has the lowest SOC exhibits the lowest open circuit voltage, thus decreasing the total battery pack voltage; the discharge algorithm would anyway continue discharging until the total series voltage reaches the lowest limit, but the cell with the lowest voltage will reach a SOC under its minimum acceptable level, thus undergoing overdischarging. Also, if cells had different SOC, some of them would not be discharged up to their lowest SOC, therefore, leaving some charge in the ESU that is not used. Analogously, during charge the cell that has the highest SOC and therefore the highest open circuit voltage will undergo overcharge because total voltage is lowered by the other cells with lower SOC and lower open circuit voltage. In such cases, the total available life of the battery pack would be limited to the life of the cell that sustains the worst overcharge or overdischarge cycle, and failure modes would occur earlier than normal. Balancing therefore intervenes by equalizing the charge in each of the cells, granting the same performance behavior from each one of the cells in the ESU (Fig. 5.6).

FIGURE 5.6
Unbalanced cells that must be managed by the cell balancing function of the BMS.

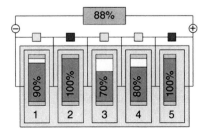

Techniques employed for cell balancing can be passive or active. In *passive balancing*, energy in cells that have high SOC is wasted as heat in resistors (method known as "resistor bleeding balancing"); in *active balancing*, a control logic computes the SOC of each cell and directs via DC–DC converters the energy flow between cells to equalize each cell's SOC.

To avoid high current spikes to flow in uncharged loads when the EMS commands the discharge of the ESU, a *precharging* circuit, using DC choppers or R-C circuits, is used. With pre-charging, no sparks would occur at the terminals of on-off contactors that connect the ESS to the load, and smooth ramping up of power can be achieved to avoid damages to the electric system.

5.5　Power Plant Controller and Power Conversion System

The PPC and the PCS are the ESS system components that manage the conversion of the energy to and from the ESU. The PPC and PCS are composed of a number of hardware and software subsystems that span from technologies like automation, power electronics, telecommunications, software, and control logic. It is outside the scope of this book to cover all the topics, but an overview of the most important concepts will be given in the next sections. PPC manages the optimal functioning of the BMS and the PCS by interfacing with the high-level commands from the EMS.

5.5.1　The Modbus Protocol

The communication between electronic devices in the ESS can be performed via many different communication protocols, but the one that is most commonly used as the de facto industrial standard is the *Modbus protocol*. The advantages underlying its wide usage are the fact that it is open source, free of any kind of payment and royalty, that it is designed to be easy to implement and maintain, and that it can link practically all kind of components, from relays to sensors, from actuators to SCADA, from PLCs to industrial PCs [1].

Modbus is designed to be capable of linking a number in the range of over 200 devices, all connected to the same bus network topology, where a *bus network* is the connection of the devices (named *nodes*) to the same communication backbone. In small networks, a bus topology has the advantages of being very simple to implement, with minimal use of cabling. Practically all devices that are relevant to industrial automation use are Modbus compliant. The Modbus protocol is maintained by a panel of independent users to make sure that the Modbus protocol is kept updated with the industrial automation community needs.

Every device in the Modbus network has a unique address, and Modbus commands are sent in a master-slave configuration by one main device, the [*master*], to individual devices (the [slaves]) with the instruction to be performed. All commands have a checksum to avoid transmission errors, and devices provide an acknowledge message to the master. Broadcast massages are also possible, to which slaves do not acknowledge reception.

The communication frame formats to be chosen from are the RTU, the ASCII, and the TCP, containing the address field, the instruction, the data to be sent, and the checksum. Slave response contains the confirmation of the action taken, the data, the checksum. In case of errors, the slave sends an error message to which the master takes appropriate counteractions.

The byte order is the *big-endian*, where, the most significant byte value is stored at the memory location with the lowest address. Data types do not cover the full range of data structures available in the software and telecommunication community, but can anyway be enough varied (from single bits to 32-bit registers, from mixed types, to standardized IEEE floating point) to provide ample industrial automation communication efficiency. Operations on data are the usual "read," "write," "mask" (for bitwise operations), and "store" on single or multiple registers.

5.5.2 The Smart Grid Protocol IEC 61850

The emergence of a new type of energy users capable not only to consume but also to generate energy, and the capabilities offered by information technology, is fostering the development of a new paradigm in the management of power in the electricity grid. Distributed generation means that energy is flowing bidirectionally, and not only one-directionally as before.

The capability to embed intelligent logic in once-dumb power devices is empowering the new concept of an internet of energy, alongside the internet of data and the internet of things. These devices, nicknamed *intelligent electronic device* (IED), range from switchgears to motor drive control boards, from PCS to transformers, capacitor banks, circuit breakers, and more devices.

To enable fine control of IED in the electricity grid, the smart-grid protocol IEC 61850 has been developed for the electrical substation automation by integrating two parallel network infrastructures: the power distribution network, and the data network. The protocol provides control, measurement, and monitoring functions by creating a standard to which smart grid devices must comply with [2]. IEC 61850 is designed to be vendor independent.

The use of a communication protocol for power devices cater for a wealth of advantages that can be summarized with the following:

- adaptability: power can be delivered to different flow paths to adapt to changing conditions (*power rerouting*)
- energy management: power quality is improved by applications like reactive power compensation, load-shedding (also known as interruptibility), use of storage
- infrastructure investment optimization: better planning for refurbishments and renewal of components and power lines
- integration: different grids from different operators can be more efficiently connected and power flows can be routed to or from new paths improving reliability of the overall system
- interactivity: customers become energy producers, can be enabled to manage their power outflows therefore disrupting the old business model of centralized utilities
- optimization: energy can be delivered to procure improved efficiency and economic performance
- ordinary maintenance: fault location and isolation, monitoring and measuring reduce grid downtime, and increase quality of service
- predictive maintenance: using data collection on physical parameters (i.e., operating temperature, currents. . .), outages can be prevented

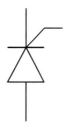

FIGURE 5.7 Symbol of the thyristor.

- smart metering and monitoring: extremely efficient and detailed data-logging is made available to grid operators and final users
- secure: the electric grid can be made more stable, more reliable, and large black-outs can be prevented by having electricity flowing in alternative paths and overcoming nodes where power is off

The control model for fast transfer of event data over the network is provided by the *generic substation events* (GSEs) model; the transfer of data occurs between the IEDs as multicast, sent to a subset of all networked IEDs, or as broadcast to all IEDs. The GSE control model uses the *generic object oriented substation events* (GOOSE) event transfer mechanism to ensure transmission speed and reliability. Data are embedded into Ethernet data packets and use the "publish-subscribe" messaging system.[2] GOOSE uses priority tagging to create different logical subnetworks inside the original IED physical network, and messages can be re-transmitted at varying intervals or on an event basis.

The *Substation Configuration Language* (SCL) is the format used for the data representation and the associated function definition of IEDs. Every IED is represented as the set of its logical devices, access points, logical nodes, the messages under the GOOSE model. The capabilities of each IED are defined as the *IED Capability Description* (ICD) file, where the ICD file has to be supplied by the IED vendor. The complete substation system containing all IEDs is described by the *System Specification Description* (SSD) file, where substations are fully described in a *Substation Configuration Description* (SCD) file.

IEC 61850 uses the *Manufacturing Message Specification* (MMS) international standard (ISO 9506), with Ethernet and TCP/IP for data transport. MMS defines a set of *virtual manufacturing device* (VMD) objects, on which the usual set of operations on data are applied, a set of standard messages to be exchanged for monitoring and control of VMDs, and a set of rules for message encoding.

5.5.3   Power Electronics

It is outside the scope of this book to provide a detailed description of the electronic components used in power conversion systems. Anyway, since some of them are frequently referred to when working with energy storage systems, some basic knowledge on them can be useful.

The use of some components have already been shown in previous chapters; namely, thyristors and insulated-bipolar gate transistors. Such components, together with many others, constitute the basis for more complex circuits that are employed to condition

---

[2] The *publish-subscribe* messaging defines senders of messages as the publishers, nodes who do not know which other nodes will be reading the messages sent, the subscribers.

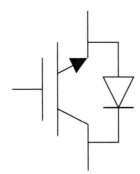

Figure 5.8 Symbol of the IGBT.

power according to the various needs of power plants. Systems like DC–DC converters and DC–AC converters are the building blocks of energy storage power conversion systems.

Thyristors and IGBT

A *thyristor* [also called *silicon controlled rectifier* (SCR)] is a switch with a conductive and nonconductive mode, built of semiconductor layers of different electrical characteristics (see Fig. 5.7). When they are controlled via an trigger current on their gate input, they switch on letting the current to flow through the device if they have a suitable voltage between the emitter and the collector (forward-bias). If the thyristor is not controlled via the gate or the voltage is reversed, the device acts as an off-switch. Figure 5.8 shows the symbol for the thyristor.

Also the *insulated-gate bipolar transistor* (IGBT) is a three-terminal power semiconductor device used as a switch when fast switching applications and high efficiency, and complex waveforms by means of sophisticated control logics are required by the field of application. Figure 5.8 shows the symbol for the IGBT.

DC–DC Converters

Since electric storage normally outputs energy as DC voltage and current, the use of electronic circuits to condition DC power are of widespread use in the PCS systems. When storage energy is discharged or charged in ESU, unregulated DC power is transformed into regulated DC power with values, and eventually polarity, different from the input DC power. One example of widespread use is the rectification of AC voltage from the grid distribution (phase-to-neutral typical AC voltages of 120 or 240 V) to DC voltage, which is then regulated to a lower voltage to power devices like personal computers or domestic appliances.

DC–DC conversion is achieved by circuits that operate in switched mode, called *DC choppers*. Since such circuits dissipate very low power, converters operating in switched mode show high to very high-efficiency conversion factors, ranging from 70 to 95% [3].

The circuit is operated, in its simplest form, by a switch that alternates between on-off positions according to a duty-cycle D in the interval [0, 1]. If the input voltage is constant, the output is a rectangular voltage wave form of period T_s (and switching frequency $f_s = 1/T_s$) with "on" time as long as DT_s, and "off" time of $D(1 - T_s)$ (Fig. 5.9). The logic that drives the functioning of the switch uses *pulse-width modulation* (PWM) to achieve the desired output voltage level.

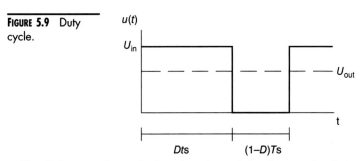

FIGURE 5.9   Duty cycle.

The *DC conversion ratio* C_{DC}, function of the duty cycle D, is the ratio between the output voltage U_{out} and the input voltage U_{in}:

$$C_{DC} = \frac{U_{out}}{U_{in}} \tag{5.1}$$

where the average value of U_{out} as a function of the duty-cycle is given by:

$$U_{out} = \int_0^{T_s} U_{in}\, dt = D\, T_s \tag{5.2}$$

A typical DC chopper circuit used when DC output voltage must be regulated to lower levels than the input DC voltage, is the *buck-converter* with conversion ratio $C_{DC} = D$. When the DC input voltage is regulated to higher levels than the input voltage, the *boost-converter* is used, characterized by a conversion ratio of $C_{DC} = 1/(1 - D)$. Both circuit schematics are shown in Fig. 5.10, where the switch can be thyristors, IGBTs, or similar components. The L-C circuit is a low-pass filter used to smooth the harmonic content of the switching operation and to pass to the load only the DC component. The *freewheeling* diode is used to provide a path for the current to flow when the switch turns off.

The two circuits can be combined in a single *buck-boost converter*, where the conversion ratio is $C_{DC} = -D/(1 - D)$.

Technical regulations almost always recommend to add a transformer to isolate output power from DC input power. Transformers are used to guarantee isolation, but since high-frequency transformers are very small if compared with low-frequency transformers, incorporating them inside the choppers can bring about low costs and low dimensions. Transformers can also be very useful when stepping up the voltage, where switching can be minimized by proper selection of the primary-to-secondary transformer ratio.

An example of a bidirectional power flow using choppers is shown in Fig. 5.11. The switches and diodes are connected in antiparallel, and the transistors are controlled to be never switched on at the same time, and, together with diodes, form current-bidirectional switches. Even the polarity of the system can be controlled by appropriate PWM on the two switches.

DC–AC Converters

The DC–AC converter, or inverter, is the electronic power circuits that convert the DC power stored in the ESU to AC power suitable for use for loads and grid injection [4]. These devices are of widespread use in many industrial, commercial, residential applications, with a history of many decades of high reliability and performance. As in DC–DC converters, PWM is the technology used in inverter control logic to obtain the desired

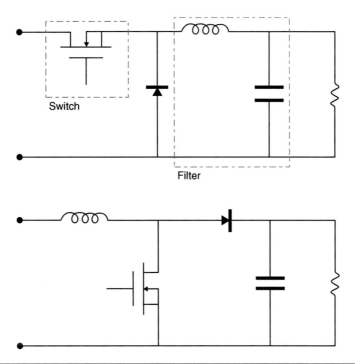

FIGURE 5.10 Schematic of the buck-converter (above), and a boost-converter (below).

output voltage and frequency. Inverters used in ESS need to be bidirectional, meaning that conversion also must occur as AC–DC alongside DC–AC. While DC–AC function is sometimes called *motoring mode*, AC–DC is called *regenerative mode*, where power flows back from the previous AC load, which now becomes a generator.

A single-phase inverter circuit example is the *full-bridge inverter* (Fig. 5.12). Since the input that is processed by the circuit is the voltage, this circuit configuration is also called *voltage-source inverter* (VSI). Four valid switch states are available for this circuit configuration, and control logic must assure that other states are not entered to avoid malfunctions, that switches do not change state simultaneously to avoid short-circuiting, and

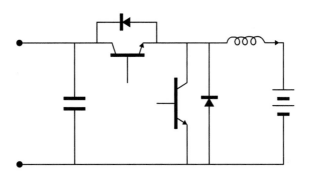

FIGURE 5.11 Bidirectional buck-converter for battery charge/discharge.

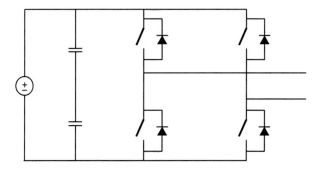

Figure 5.12 Single-phase full-bridge inverter circuit.

correct amplitude, phase and frequency is output to the load or to the grid. A three-phase inverter schematic is like the full-bridge inverter but with two additional branches in parallel to cater for the two additional phases of the three-phase system. Eight valid switch states are available for this circuit configuration, and control logic must assure respect of the same conditions as per single-phase full bridge VSI (Fig. 5.12 and Figure 5.13 for the three-phase version). Due to the switching nature of the circuit, filtering is needed at the output to avoid the injection of harmonics in the load or the grid.

Apparent power A is given by:

$$A = \sqrt{P^2 + Q^2} = V \times I \tag{5.3}$$

where P is the active power, Q is the reactive power, and V the voltage and I the current. By squaring:

$$A^2 = P^2 + Q^2 = V^2 \times I^2 \tag{5.4}$$

that can be considered as a circle on the $I-V$ plane, with center at origin and radius equal to $V \times I$. The circle can be divided in four quadrants that are numbered as:

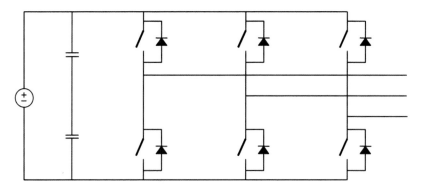

Figure 5.13 Three-phase full-bridge VSI circuit.

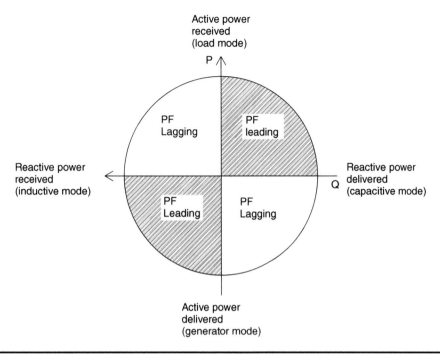

FIGURE 5.14 Inverter bidirectional active and reactive power capability curve.

- quadrant 1: $V > 0, I > 0$
- quadrant 2: $V > 0, I < 0$
- quadrant 3: $V < 0, I < 0$
- quadrant 4: $V < 0, I > 0$

In quadrants 1 and 3, power flow is positive and the inverter is providing energy to the load (motoring mode); in quadrants 2 and 4 power flow is negative, and the inverter is absorbing energy from the supply (regenerative mode).

Inverters used in ESS are required by regulations to be capable of providing reactive power to the load or the grid. Reactive power capability is needed to provide grid transient stability, fault reliability, voltage rise, sag or fluctuation mitigation, especially with the increase of penetration of distributed generation and renewable energy decentralized power plants [5]. Static converters like DC–AC converters have the ability to deliver or receive active and reactive power over the area of the circle $P-Q$ shown in Fig. 5.14, where:

- *lagging power* flow occurs when the real and reactive currents flow in the same directions, as in the case of an inductive load receiving both real and reactive power, or a capacitive generator delivering both real and reactive power.
- *leading power* flow occurs when the real and reactive currents flow in opposite directions, as in the case of a capacitive load receiving real power and providing reactive power, or an inductive generator providing real power and receiving reactive power.

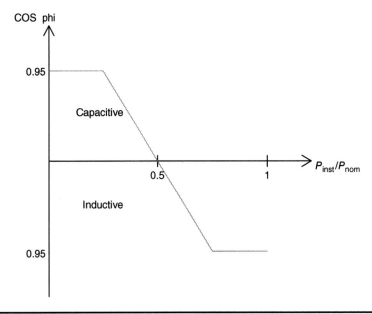

FIGURE 5.15 Example of a characteristic curve for reactive power.

Grid codes from utilities or national regulators can provide the characteristic curve for the injection of reactive power in the grid depending on grid variables like, for example, line voltage. For instance, Fig. 5.15 show the specification for reactive power as a function of the ratio between instantaneous power P_{inst} and nominal power P_{nom}. When the ratio is under 50%, the generator is required to operate in capacitive mode; when the ratio is over 50%, the generator is required to operate in inductive mode.

Regulations also provide the specifications for the capability curves to which the inverter in a ESS must comply with to receive connection permitting. Figure 5.16 shows two possible capability curves: one triangular and a second rectangular. In case of electrochemical energy storage, the capability is the combination of the inverter capability (circular) with the battery capability (rectangular). The ESS capability curve will therefore need to comply with the limits set forth for absorption or injection of reactive and active power in the capability curves set forth in the grid code.

All inverters have minimum, maximum current and voltage requirements, whereas the highest conversion efficiency is reached when the converter operates at its highest possible voltage and current. Normally, the nominal AC voltage of the converter has to be lower than 2/3 of the minimum DC battery voltage to operate properly; this is why the choice of the minimum DC battery voltage affects the power rating of the converter. Of course, the converter output voltage is not relevant since a transformer on the AC side can increase or decrease voltage of the bus that is connected to the grid or the load. The choice of the battery voltage has therefore a large impact on the cost and power of the PCS and the overall energy storage system.

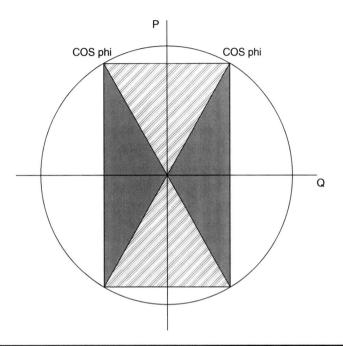

FIGURE 5.16   Example of a capability curve of an ESS.

5.6   Grid Interfacing and System Functioning

The interfacing and functioning of the ESS in the distribution grid depends on technical norms and regulations. Typically, an ESS is considered as a generator, not a load, and has to be connected to the grid after having obtained a connection permit and having received the technical specifications for the connection. Both connection permit and technical specifications depend on the characteristics of the point of connection and the grid.

All values that can be described in this section are indicative; local grid codes (i.e., German BDEW, UK National Grid code, US FERC, or the Italian Terna's Codice di Rete) have to be checked to ensure compliance of ESS design.

5.6.1   Interface Protection System

An ESS is defined as an *active connection*, as opposed to *passive connection* of plants the only discharge energy from the grid to power their connected loads. Typically, all active connections on a distribution grid must be equipped with an anti-islanding interface protection device that prevents the active connection to inject energy in the grid when a grid fault occurs. *Islanding* is indeed the condition during which a generator continues to power a load even if such load is disconnected from the grid. In general terms, when a generator is connected in parallel with the load and the distribution grid (i.e., the generator can provide power to both the load and the grid), if the grid goes down no generators out of control of the grid operator are allowed to inject energy in the grid itself. The main reason is to avoid safety issues with the operators working on the power lines in case of troubleshooting or maintenance. Therefore, apart from normal protection and command switches that are compulsory from technical regulations and state-of-the-art electrical technology, an *interface Protection System* (IPS) must be installed in the ESS to prevent its islanding if the ESS is connected in parallel to the grid.

Typically, the functioning of an ESS that is connected to the distribution grid must:

- not cause disturbances to the normal functioning of the grid
- preserve the quality of service of the grid
- disconnect with no intentional delays when the grid is absent
- disconnect when grid voltage and frequency levels are outside specified values
- disconnect if the IPS is in a faulty condition

The IPS comprises of a set of electronic relays that, when certain parameters are recognized not to belong to specified ranges, drive a switch open to disconnect the ESS from the grid. The parameters that are checked by the logic depend by local regulations and the type of grid voltage (LV, MV, or HV). An example of values employed for interface protection are provided in Table 5.2.

For each relay, more than one set of values can be given. Typically, two sets are available: one restrictive, the other permissive, depending on fault conditions. For power

Grid: LV			
Parameter	**Relay**	**Delay**	**Value**
Maximum voltage (mobile ave. 10 m)	59	max 600 s	1.1 U_n
Maximum voltage	59	200 ms	1.15 U_n
Minimum voltage	27	400 ms	$\leq 0.85\ U_n$
Maximum frequency	81>	100 ms	+1% of nominal
Minimum frequency	81<	100 ms	−1% of nominal
Grid: MV/HV			
Parameter	**Relay**	**Delay**	**Value**
Maximum voltage	59	max 600 s	1.1 U_n
Minimum voltage	27	1500 ms	$\leq 0.85\ U_n$
Maximum frequency	81>	150 ms	+0.4% of nominal
Minimum frequency	81<	150 ms	−0.4% of nominal
Maximum residual voltage	59V0	25 s	5% of nominal
Maximum homopolar voltage	59N	3 s	≥ 10 V
Maximum voltage inverse sequence	59 Vi	no delay	15%
Minimum voltage direct sequence	59 Vd	no delay	70% U_n/E_n

TABLE 5.2 Example of Parameter Values for Interface Protection. Values are Given for Reference Only Since they Depend on Local Regulations

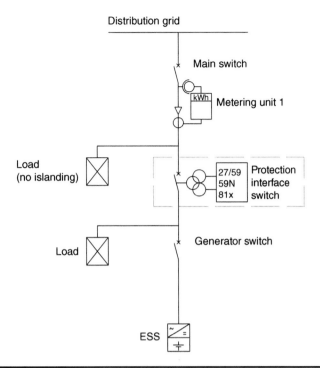

FIGURE 5.17 Example of an ESS single-line schematic showing main, generator, and interface protection switches.

plants over a certain power threshold and depending on voltage level (i.e., over 20 kVA for LV, over 400 kVA for MV), a redundancy relay is added to the protection system to provide additional safety.

The insertion of the protection switch in a single-line schematic of an ESS is demonstrated in Fig. 5.17. In the case shown, one of the loads is still functioning in island-mode if the interface protection disconnects the ESS from the grid, but the IPS prevents the active component (the ESS) from injecting energy into the grid.

According to IEC 61850, the IPS device can be connected to the internet to allow for the reception of a disconnect signal from the grid operator, or to change the relays pre-set values.

Normally, ESS that are working in parallel with the distribution grid must maintain a voltage level within the 85%–110% range and a frequency range within ±5% of the nominal frequency. ESS must also avoid injecting in the AC grid, currents with DC components higher than 0.5%; this requisite can be achieved using a transformer at the grid frequency, or a protection switch that is capable of sensing the injection of DC currents. If the DC current overcomes the 1 A limit, the protection must switch off the generator with timings in the range of 200 ms, or 1 s if the DC current goes over 0.5% of the inverter nominal current.

5.6.2 Functioning and Grid Services

Normally, ESS connect to the load and the distribution grid by means of static converters, namely, the inverters that provide the DC to AV conversion. Inverters must respect local regulations that typically require them to respect ranges of operations for their main parameters. For instance, voltage levels must be within 85–110% of the nominal values; synchronization must normally be within a specified frequency range (e.g., 49.90–50.10 Hz for some European countries); voltage and frequency must be stable within the thresholds for a specified period of time (in the range of 30–300 s, depending on grid codes, and if connection occurs as startup or after a fault); power injection must occur gradually and linearly, for example, with a gradient as high as 20% of the maximum power of the generator.

The functioning of generators and ESS entails the following configurations:

- short period grid-connected functioning: is typically allowed for generators, if parallel mode does not exceed a time frame in the range of 10–30 s
- full off-grid functioning (islanding): an IPS must be installed and comply with local regulations
- continuous grid-connected functioning: must comply with the local regulations set forth in national or operators' grid codes, in particular by respecting voltage, frequency, and grid service specifications

To avoid disconnection during voltage sags or spikes, the ESS must comply with the *fault ride through* (FRT) functional specifications. The generator that complies with FRT remains connected to the grid in case grid voltage levels decrease or increase over nominal up and low range limits and for specified periods of time. The rationale is that disconnecting generators during temporary grid faults would cause more damages than provide benefits to grid stability.

When FRT is for undervoltage fault conditions, the *low-voltage ride through* (LVRT) curve applies. Shown in Fig. 5.18 is an example of a LVRT curve: if the phase-to-phase voltage decreases under the nominal low level for the specified amount of time, the generator must remain connected to the grid, and active power injection is permitted; under the curve, generator must disconnect.

A similar behavior is permitted in the case grid voltage levels overcome the upper voltage limit; in this case, *over-voltage ride through* curve specifications must be complied with by connected generators.

The ESS is required by regulations to be capable of providing a number of services to the distribution grid which are:

- active power regulation
- voltage control
- voltage support during short-circuit
- frequency tolerance and load-shedding

Active Power Regulation

Active power must be controlled and regulated by the generator. Active power must be limited if voltage levels are close to their upper limit (typically 110% of nominal);

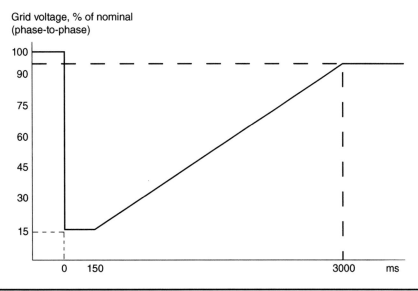

Grid voltage, % of nominal
(phase-to-phase)

FIGURE 5.18 Example of a LVRT curve.

reduced if over-frequency transients occur in the grid; increased if underfrequency transients occur in the grid.

Reduction or increase of active power injection must comply with frequency–power response curves that depend from local regulations. Normally, regulation must occur gradually and linearly with response times under a couple of seconds and specified gradients. Active power injection will not be increased back to original level until grid frequency returns within the nominal range, at least for a time period in the range of 300 s. Active power limitation can also be requested to occur automatically after a remote signal from the grid operator.

Voltage Control

Distributed generators are potentially capable of raising the grid voltage level over regulated voltage limits. Normally, the average RMS voltage value must not be higher than 110% of U_n for more than ten minutes. To respect such limits, typically for values of grid voltage over 120% of U_n over a certain time frame (i.e., 0.6–0.7 s), the generator must disconnect from the grid. When the average voltage over a time frame of, for instance, 10 min and the mobile average surpasses 110% of U_n, the generator disconnects within a time frame of a few seconds (i.e., 3 s). These operations are enforced by the use of the IPS as discussed in Sec. 5.6.1.

For example, distributed REN power plant influence voltage levels when connected to the distribution lines especially where feeders are weak and undersized. If the REN power plant is connected close to the voltage substation, the compensation occurring in the substation itself can run the risk of reducing the value of the voltage at the end of the feeder line. If the REN plant is connected to the end of the feeder line, the local voltage can be raised to levels higher than the upper threshold. The impact can be mitigated if the generator can manage reactive power flow and not only active power. Normally, generators work with power factors at $\cos \phi = 1$; functioning to power factors different from 1 can be required by regulations.

Voltage Support during Short-Circuit
During grid short-circuit events, the generator must provide voltage support by injecting reactive power to avoid voltage sags. The typical power factor for generators is ± 0.95, where the reactive power range can be adjusted continuously within this range depending on local requirements for reactive power grid injection.

Frequency Tolerance and Load-Shedding
While voltage variations effects are local, instability of frequency have an effect on the entire network. The ESS, therefore, is required to be disconnected in case of need, where generator disconnection is typically coordinated under nation-wide *under-frequency load shedding* schemes. Disconnection can be requested to occur with no delay (*fast disconnection mode*) or after a time frame (*slow disconnection mode*) specified by regulations, depending on the type of emergency that the network is facing. Fast disconnection is employed for frequency control or faults that seriously hamper grid stability. Slow disconnection is used to manage overvoltage, insufficient supply, or grid congestion that cannot be taken care of otherwise. Disconnections are automatic and controlled via remote signal, and disconnections can be on the partial or the total quota of generation power, where slow disconnect can also be accepted as being manually controlled.

5.7 Load Analysis
Depending on the applications for the ESS, one of the first steps in system design is the sizing of the load. In case the load is the distribution grid, the feeder line is considered as an infinite load or generator that can absorb or supply an infinite amount of electric energy. In such case, no sizing at all is needed and all ESS needs can be fulfilled when the ESS is connected to the distribution grid. In case the load is not connected to the grid, the system is said to be in *island* or in *stand-alone mode*. If this is the case, the sizing of the ESS must consider the way the load operates and must be capable of assisting it with no or marginal *capacity shortage*, namely the positive difference between load energy demand and ESS energy supply. A good knowledge of the load is therefore mandatory for good design.

Load can be collected or estimated as the series of power data at a suitable time granularity (from seconds to hours) in basically three ways:

- ex-ante (or forecast) data analysis
- ex-post (or historical) data analysis
- estimation

Ex-ante (forecast) load data analysis is the use of data that represent the power demand in the following days as forecast by the analyst or as provided by the final user (i.e., the operations manager of an industrial site) with a suitable time granularity. Such forecasts can be used as input to algorithms in order to find the ESS set-points which will be used to control the ESS operations according to the constraints and utility functions that are requested by the specificity of the selected applications. For instance, a time-of-use application will use the forecast data to schedule the best timing for ESS charge and discharge according to industrial load and electricity prices along the 24 h of the day.

Ex-post load data analysis is the collection of the power demand as historically logged data by the user in the periods preceding the analysis. An example of a load logged from an industrial load can be found in Fig. 5.19. An analysis of the load can provide a lot of insights on which exact applications can be provided by the ESS to the industrial customer.

Load estimation is an algorithmic way to compute the likely load needs that certain users are supposed to be needing depending on their characteristics. There exist different ways to estimate the load; for instance, if the number and characteristics of each connected loads are known, the equation that can be used to estimate the energy need is the following:

$$E_{\text{load}} = \sum_{i=1}^{N} E_k \tag{5.5}$$

where E_k is the energy needed, or absorbed during operations, for load k, given by:

$$E_i = P_i \cdot h_i \cdot K_{c,i} \cdot K_{u,i} \quad \forall i = 1, N \tag{5.6}$$

where P_i is the nominal power of each load, h_i is the number of hours during the day when the load is absorbing energy (namely, is in operation), $K_{c,i}$ is the *coincidence factor* which represents the correction coefficient (ranging from 0 to 1) that takes into account simultaneity of functioning for each load, and $K_{u,i}$ is the *usage factor* which represents the correction coefficient (ranging from 0 to 1) that is to be applied to account for power draw of load i lower than its nominal power. Technical norms and regulations provide values for K_c and K_u that can be used by the designer to estimate the total load.

5.8   Electricity Price Analysis

Depending on the application, knowledge of the electricity prices is fundamental for proper business planning and power purchase agreement contracts.

Electricity market organization can be very different across countries, but within a single country electricity prices are normally formed in the *spot market* (the other being the *derivative* market) comprising of three electric energy markets over different time frames:

- the *day-ahead market*, where producers, wholesalers and eligible final customers may sell or buy electricity for the following day.

- the *intra-day market*, or *balancing market*, where producers, wholesalers and final customers may modify the injection/withdrawal schedules that they have proposed in the day-ahead market.

- the *ancillary services market*, where the grid operators procure the resources necessary for managing, operating, monitoring and controlling the power system (i.e., the relief of intrazonal congestion, the creation of energy reserve, the real-time balancing).

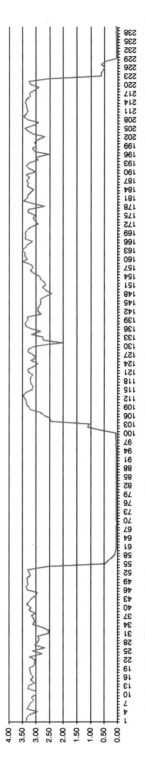

FIGURE 5.19 Example of a power load (in MW) logged at an industrial site over a period of 10 days.

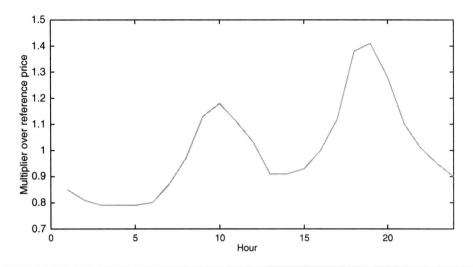

FIGURE 5.20 Example of intraday electricity price variations.

The majority of the transactions takes place during the day-ahead market, where energy is traded for each hour of the following day. Participants submit offers or bids with the quantity and the minimum or maximum price at which they are willing to sell or purchase. Bids and offers are accepted under the economic merit order criterion[3] while taking into account the transmission capacity limits between different areas in the country. The day-ahead is an auction-based market where the price is determined by the hourly intersection of the demand and supply curves and is differentiated from zone to zone when transmission capacity limits are saturated. The merit order ranks offers according to their ascending price (which is a function of the marginal costs of energy production), and matches the amount of energy offered with the amount of energy demanded. The offers with the lowest marginal costs are the first ones to be injected in the grid to meet demand. As a consequence, the merit order criterion is intended to provide a competitive advantage to the sources of energy that have the lowest marginal costs of production, therefore managing to minimize the electricity costs.

An example of price trend for an average day is given in Fig. 5.20. In energy markets where the penetration of REN are high, electricity price daily trends show a double daily peak with a decrease during the central hours of the day.

Since storage grants the user the possibility to time-shift, the delivery of energy to the grid, ESS can extract more value from the power plant by injecting electricity in the grid when the electricity prices are higher. Injecting energy during central hours would cause reduction in revenue streams, while an improved control logic will avoid sale during off-peak central hours by shifting energy injection to peak hours. Since prices are also highly correlated with power draw from distribution lines, selling electricity during peak

---

[3] The *merit order* is the price formation mechanism that matches demand and supply by giving sale priority to the least expensive quantities of a product, and gradually bring into the market other quantities of the product with marginally higher prices until demand is fully satisfied. In this way, the total supply has the lowest weighted average cost.

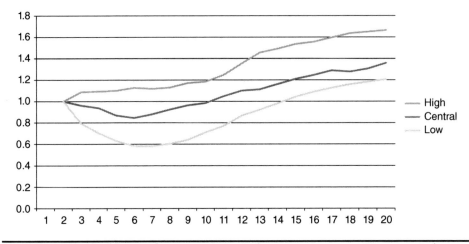

FIGURE 5.21 Example of forecast for price variation in three scenarios over a 20 year period.

hours would improve, as a side-effect, the electrical stability of the electricity distribution network.

Energy prices are forecast by specialized advisory firms to provide decision-makers with reasonably accurate figures to plan energy investment over the long-term period (see Fig. 5.21). A number of different scenarios, based on different assumptions on macroeconomics conditions, on new addition or divestment in the electricity generation park, on planned and current investments in electricity grid refurbishment or expansion, are taken into account in advanced modeling algorithms for electricity price forecasting.

5.9   Optimization Methods

Optimization techniques are of strategic importance for all energy projects. The goal of optimization is the maximization or minimization of the utility function chosen by the organization. Such function is normally linked to the highest possible energy conversion and its availability for use, which translates into the maximization of the revenues of the project and the minimization of its costs. A correct implementation of these techniques makes the difference between a go or a no-go decision to project development and construction.

From the Oxford English Dictionary, optimization is defined as [6]:

> . . . the action or process of making the best of something; the action or process of rendering optimal; the state or condition of being optimal.

An optimization process, therefore, is the action on a set of input variables that impacts the outcome of an *objective function* that is to be optimized. This means searching for the set of input values that fulfills the purpose of the optimization goal, whether it be maximization or minimization. When the function to be optimized is only one, the problem is a *single-objective optimization*; when multiple objective functions are considered for optimization, the problem is a *multiobjective optimization*. If the objective function must respect a set of constraints on the input variables, the problem is said to be *constrained*.

In mathematical notation, a problem of single-objective function optimization can be framed as follows [7]. Given a vector x in a euclidean space $\mathbb{R}^n$ of the input variables, a vector w in a euclidean space $\mathbb{R}^p$ of the output parameters, a scalar y in a euclidean space $\mathbb{R}$ of the output variables, spaces $\mathbb{R}^n$, $\mathbb{R}^p$, and $\mathbb{R}$ can be restricted by the set of constraints to the subset spaces $X \subseteq \mathbb{R}^n$ with $x \in X$, $W \subseteq \mathbb{R}^p$ with $w \in W$, and $Y \subseteq \mathbb{R}$ with $y \in Y$. X is called the *design space*, and Y is the *solution space*.

The following functions can be defined on X:

$$g(x) : X \subseteq \mathbb{R}^n \to W \in \mathbb{R}^p, \quad w = g(x) \tag{5.7}$$

where g is the function that defines the output parameters, and

$$f(x) : X \subseteq \mathbb{R}^n \to Y \in \mathbb{R}, \quad y = f(x, w) = f(x, g(x)) = f(x) \tag{5.8}$$

where f is the single-objective function. Single optimization is obtained by solving the following problem:

$$\text{optimize } f(x, w), \quad x \in X, \, w \in W \tag{5.9}$$

Multiobjective optimization problems concern the search of the best solution for more than just one objective function. Optimization therefore concerns the vector $f(x)$ defined as:

$$f(x) = \left(f_1(x), \ldots, f_n(x) \right), \quad x \in X \tag{5.10}$$

where $f_i \, \forall i = 1, \ldots, n$ are the objective functions of x in the design space.

In case of multiobjective optimization, due to conflicting conditions of optimality, there is very likely no single solution capable of optimizing all the objective functions at the same time. The condition of *Pareto optimality* will occur when, if changing one of the vectors in the design space, at least one of the objective functions will yield a result that will cause such function to obtain a value which is worse than the one previously attained.

A vector $f(x_1)$ is said to *dominate* a vector $f(x_2)$ if and only if $f_i(x_1) \leq f_i(x_2) \, \forall i$ and at least a j exists so that $f_j(x_1) < f_j(x_2)$. A point x^* in the design space is *Pareto optimal* if the vector of the objective functions $y = f(x^*)$ is *nondominated* in the solution space.

The *Pareto Frontier* is the locus of the objective functions $f(x)$ in the solution space for which the Pareto optimality condition holds true, that is where the vectors $\{f\}$ are non-dominated. The vector $\left(x_1^*, \ldots, x_n^* \right)$ in the solution space at Pareto optimality is the vector of the *Pareto-optimal solutions*.

The Pareto frontier is not the optimal solution of each single-objective function per se, but rather a set of the various possible values that the objective functions can attain by choosing between the input variables belonging to the Pareto optimal solutions. What this means is that the designer can choose the trade-off that suits the overall problem best. One advantage of finding the Pareto frontier is that if the decision maker wants, in later moment, a different trade-off between all the objective functions, there is no need

FIGURE 5.22
Bidimensional
Pareto frontier
reached after
different numbers
of iterations of a
multi-objective
optimization
algorithm (modified
from [8]).

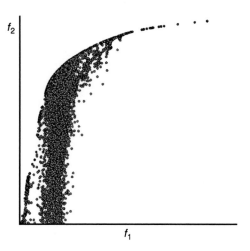

to run a new computation because the set of solutions has already been found, and it is just a matter of selecting the $x \in X$ that provides the new desired configuration of the objective functions.

Figure 5.22 shows an example of a Pareto frontier reached by a multiobjective algorithm after a growing number of iterations.

Different algorithms can be employed to solve the optimization problem, employing a deterministic or stochastic methodologies.

5.9.1   Deterministic Optimization

Deterministic Optimization takes advantage of mathematical programming and linear algebra normally by calculating first or second derivatives of functions to achieve a single unequivocal and replicable solution. Deterministic optimization algorithms are based on rigorous mathematical formulation, and while capable of a fast convergence to a solution, such solution can be a local optimum and not necessarily a global optimum. Furthermore, deterministic optimization is applicable to single-objective optimization rather than multiobjective optimization problems.

Deterministic algorithms can be used for, and hence categorized into, two main categories: unconstrained and constrained optimization for linear[4] and nonlinear functions.

---

[4] A function $f(x)$ is linear if $f(x + y) = f(x) + f(y)$ (additivity) and $f(\alpha x) = \alpha f(x)$, $\forall \alpha \in \mathbb{R}$ (homogeneity). The additivity and homogeneity properties taken together are known as the *superposition principle*.

An *unconstrained optimization* algorithm can be generalized as a search for the optimal solution along a line of points x belonging to a design space D^n that converges to a solution x^*:

$$x = x' + \alpha s, \ \forall \alpha \in \mathbb{R} \tag{5.11}$$

where $x' \in D$ and s is the direction, and the function to be optimized $f(x)$ is assumed to be of a sufficiently smooth class C^1 or C^2 (f is continuous, differentiable and with continuous derivatives of order 1 or 2). Therefore the unconstrained optimization problem is simply to:

$$\text{optimize } f(x), \ x \in \mathbb{R}^n \tag{5.12}$$

Some algorithms to solve unconstrained optimization problems are the following:

- Simplex nonlinear programming
- Newton's and Quasi-Newton's
- Conjugate direction
- Levenberg–Marquardt

A *constrained optimization* problem is framed as:

$$\text{optimize } f(x), \ x \in \mathbb{R}^n \tag{5.13}$$

$$\text{subject to :}$$

$$c_i(x) = 0, \ i \in E \tag{5.14}$$

$$c_i(x) \leq 0, \ i \in I \tag{5.15}$$

where $f(x)$ is the objective function, $c_i(x)$ are the constraint functions, E is the space of the equality constraints, and I is the space of the inequality constraints. A vector x that respects all constraints is said to be a *feasible point* in the region $E \cap I$. If $f(x)$ and $c_i(x)$ are continuous functions in $\mathbb{R}^k$ and the *feasible region* (or the *feasible set*) is closed and nonempty, there exists a vector x^* that is a solution to problem (5.13).

If all constraints are equalities, the simplest way to solve problem (5.13) is by *elimination*, using constraint functions to solve for variables in the objective functions.

Between the algorithms to solve constrained optimization problems, depending on whether the objective function and the constraints are linear, quadratic, nonlinear, smooth,[5] nonsmooth, are:

- linear programming
- quadratic programming
- nonlinear programming

---

[5] A *smooth function* is a C^∞ function differentiable to the ∞ order.

- mixed-integer programming
- nonsmooth optimization

Many times, the optimization problem is linear and the following methods are the ones most often used to reach an optimal solution.

The *simplex algorithm* (SA) is one of the first methods devised, and is centered on the idea that the feasible set can be geometrically considered equivalent to a polytope, which is searched by the simplex method by testing a sequence of its vertices in turn until the best vertices is found which optimizes the objective function.

The *interior-point algorithm* reaches the boundary of the feasible set only after subsequent steps all taken inside the feasible region, since these methods require all iterations to strictly satisfy the inequality constraints in the problem.

A difference between simplex and interior-point methods lies in the fact that simplex methods tend to run a large number of iterations with small progresses towards the solution, while interior-point method iterations are expensive to compute but are very efficient in each iteration in getting closer to the optimal solution.

The *active-set algorithm* is a method to solve the generic optimization problem (5.13) in the following form:

$$\text{optimize } f(x), \ x \in \mathbb{R}^n \tag{5.16}$$

$$\text{subject to :}$$

$$c_i(x) \leq 0 \quad i \in I \tag{5.17}$$

where (5.17) define the feasible region, and a constraint $c_i(x)$ is *active* at x when $c_i(x) = 0$, and *inactive* at (x) when $c_i(x) < 0$, with equality constraints that are always active by definition. The *active set* is the set of all the constraints that are active at x [9]. In LP, this set represents the hyperplanes that intersect all at the solution vector. In other types of optimization problems, the active set reduces the search space for the solution by providing a subset of inequalities that are used for the search thereby reducing the complexity of the optimization problem.

An algorithm for the active-set can be devised as the following:

```
assign a feasible starting point
repeat until the optimal solution is found
    solve the equality constraints defined by the active set
    compute the Lagrange multipliers of the active set
    remove the subset of the constraints with negative Lagrange multipliers
    eliminate the infeasible constraints
end repeat
```

where the *Lagrange multipliers* are used to find local maxima and minima for the objective function f, subject to the equality constraint function $g(x, y) = c$, where f and g are at least C^1 and where the Lagrange function is given by:

$$\Lambda(x, y, \lambda) = f(x, y) + \lambda(g(x, y) - c) \tag{5.18}$$

where λ is the Lagrange multiplier.

5.9.2   Stochastic Optimization

Stochastic optimization entails the use of randomness in the search process. The following families of algorithms are typically used:

- *simulated annealing* (SA), using the concept of annealing employed in metallurgy for the heat treatment of steel.

- *game theory* (GT), using concepts from Nash's game theory on the confrontation between players trying to achieve the best outcomes in their own best interests.

- *particle swarm optimization* (PSO), where the search is performed in analogy with the movement in space of a flock of birds.

- *evolutionary algorithms* (EA) or *genetic algorithms* (GA), applying ideas from Darwin's theory of the evolution of species or the findings of genetics.

Stochastic optimization is advantageous since it grants a better search for global optima overcoming the local minima (a property called *robustness*) in case a search space is composed of many local suboptimal points. Stochastic algorithms can address true multiobjective optimization problems much more efficiently than deterministic optimization, although this can come at the expenses of a slower convergence than the one provided by deterministic optimization.

At the current stage of development, GA are the algorithms that have been proven to be very effective on a large set of optimization problems, capable of finding good solutions in a relatively short amount of time.

A genetic algorithm, considered a subclass of evolutionary algorithms, can be used to find pseudo-optimal solutions for problems with a high number of variables and non-linearities. GA can, therefore, be particularly suited to energy storage problems when storage is coupled with other energy systems in hybrid configurations [10].

GA programming starts with the definition of a set of variables that represent the *genes* (or *genotype*) encoded in *individuals* (or *phenotypes*). Individuals are initially generated randomly and are allowed to generate new individuals by a set of techniques similar to *inheritance* performed by a *crossover* mechanism between the genes of individuals giving birth to the new individual (see Fig. 5.23) and *mutation* of the genotype of such individual (see Fig. 5.24).

For each new *generation* of individuals, the *fitness* of every individual in the population is checked against predetermined criteria, and only the fit individuals are, deterministically or randomly, selected and allowed to generate new individuals to form a new population until a defined *level of fitness* has been reached. The *reproduction* and *selection* steps of the GA are therefore the two most representative blocks of a GA. The individual that will be selected along the evolution of the individuals constituting the community and that develop themselves in subsequent generations across a period of time, will contain the set of values of the solutions, or quasi-solution, to the optimization problem [11].

With these algorithms a single optimal solution is not necessarily always reachable, thus, the number of generations are often capped to save computing time. GA can be successfully used in combination with deterministic algorithms to produce an initial set of potential solutions [12] that will be refined further by the deterministic method. One of the main advantages of GA are the low computational requirements that can yield good solutions in a reasonable time-frame.

One-point crossover
Cross-over point
Parent A 1 1 0 0 1 1 0 1 0 1
Parent B 0 1 1 0 0 1 0 0 0 1
Offspring A 0 1 1 0 1 1 0 1 0 1
Offspring B 1 1 0 0 0 1 0 0 0 1
Two-point crossover
Cross-over points
Parent A 1 1 0 0 1 1 0 1 0 1
Parent B 0 1 1 0 0 1 0 0 0 1
Offspring A 0 1 1 0 1 1 0 0 0 1
Offspring B 1 1 0 0 0 1 0 1 0 1
Uniform crossover
Parent A 1 1 0 0 1 1 0 1 0 1
Parent B 0 1 1 0 0 1 0 0 0 1
Swap mask 1 0 0 1 1 0 0 1 0 1
Offspring A 0 1 0 0 0 1 0 0 0 1
Offspring B 1 1 1 0 1 1 0 1 0 1

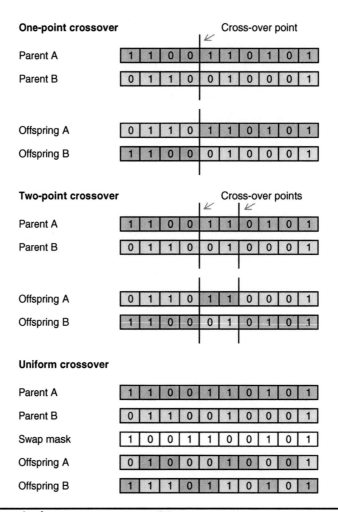

FIGURE 5.23 Example of cross-over operators in GA.

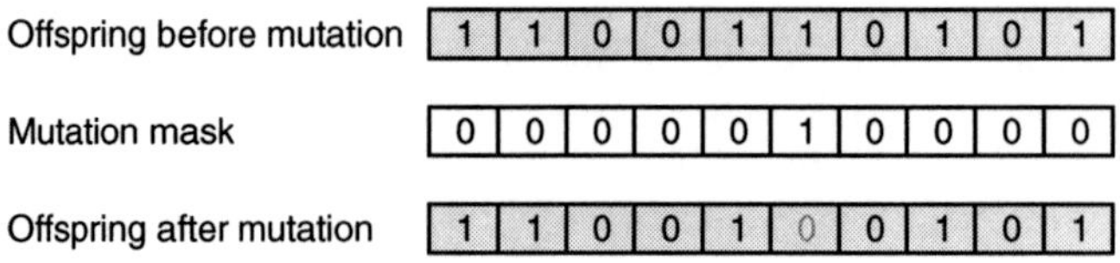

Offspring before mutation 1 1 0 0 1 1 0 1 0 1
Mutation mask 0 0 0 0 0 1 0 0 0 0
Offspring after mutation 1 1 0 0 1 0 0 1 0 1

FIGURE 5.24 Example of mutation operator in GA.

Electric energy grid services
Energy time-shift
Load following
(Up, down) regulation
Electric supply reserve capacity
Voltage support
Electric power grid infrastructure services
T&D support
T&D congestion relief
T&D upgrade avoidance/deferral
Substation on-site power
End-user energy management
Time-of-use cost management
Demand charge management (peak shaving)
Power quality and reliability
Renewable energy management
Injection profiling
Injection smoothing

TABLE 5.3   Summary of ESS Applications

5.10   ESS Design

Decision on the design of the ESS depends on the technology of the ESS and on the application, or applications, for which the ESS is to be used for, as for instance the many possible applications already outlined in the introduction of this book and summarized in Table 5.3.

The design and sizing of the ESS depends strongly on the electricity generation typology. It is outside the scope of this book to provide the reader with details on how to obtain the production estimates per hour, month, and year for solar PV, CHP, wind, or other energy conversion technologies. The reader is referred to other literature for further deepening of the subject.

In this section, a number of different applications combined with load and price information will be used to provide examples of design and sizing. Not all combinations of technologies and applications will be discussed, but the elasticity and flexibility of energy storage systems make the adaptation to different situations relatively easy. The ideas on the design and sizing given in this section should be sufficient to help the reader with the design of projects different to the ones outlined here.

5.10.1   Basic Sizing

Depending on the energy needs, a basic sizing for the ESS can be devised as the following:

$$Q = \frac{E}{V \cdot (\text{SOC}_{\text{max}} - \text{SOC}_{\text{min}}) \cdot \sqrt{\eta_{\text{RT}}}} \tag{5.19}$$

where Q is the capacity of the energy storage system (in Ah), E is the energy demand (in Wh), V is the system voltage, ($SOC_{max} - SOC_{min}$) is the available percentage of energy retrievable from the energy storage system, and η_{RT} is the RT efficiency taking into consideration all losses during a charge–discharge cycle.

Depending on ESS technology, it could be necessary that a number of storage cells be connected in series to achieve the desired voltage to be provided to the system bus-bar, and that a number of such series be connected in parallel to achieve the amount of stored energy required.

5.10.2   Simplified Sizing with Renewable Energy

ESS can be coupled with a solar PV plant for the use of PV energy to assist an industrial load. To perform sizing of the system, hourly data of PV production and load needs must be collected or estimated over a period of at least one year; the differences in hourly power between the corresponding PV energy production and load demand are used to determine the battery energy, while battery power depends on the load power that the ESS has to serve. The ESS therefore performs a time-of-use application, storing excess REN energy and using it when solar PV energy is not sufficient or is not available.

Figure 5.25 shows the hourly data for the typical day of the month for each of the months. These values can be computed using specialized software for PV yield evaluation or can be measured by energy analyzers and data-loggers.

Load power can be collected from the customer's records or budgeted with the operations managers of the plant. Figure 5.26 shows a possible sheet recording the load power values needed during plant operations.

The difference between PV power supply and load power demand is then computed and sizing is performed either on the average values for the year, or for the month during which the difference between PV power and load is the highest. In case sizing is performed on yearly average values as in Fig. 5.27, the energy that is chosen to be stored in the ESS results from the decision to charge when differences between PV and load power are positive.

Charging and discharging occur during a number of hours that depends on ESS C-rate (in this case, C/2). The energy that can be stored by the ESS is 400 kWh, the sum of the positive differences between PV power supply and load power demand adjusted for the minimum SOC and RT efficiency. The graph shows the energy content in the ESS that has, as a lower limit, the minimum SOC, and the power that is used to charge and discharge the ESS.

Similarly, the sizing is repeated exactly as in the previous case for the month with the highest difference between PV power supply and load power demand. Figure 5.28 shows the changes in sizing due to difference between yearly averages and monthly highest values.

5.10.3   Optimized Load-Based Design and Dynamic Simulation with Model-Based Design Software

Building on the concepts set forth in Sec. 5.7 on load analysis, Sec. 5.8 on electricity price analysis, and Sec. 5.9 on optimization methods, it is possible to construct an objective function that optimizes the use of ESS and provides the SPs that can be used by the EMS to extract the best economics from the energy storage project.

PV

[kW]

Hour \ Month	1	2	3	4	5	6	7	8	9	10	11	12	AVE
0	–	–	–	–	–	–	–	–	–	–	–	–	–
1	–	–	–	–	–	–	–	–	–	–	–	–	–
2	–	–	–	–	–	–	–	–	–	–	–	–	–
3	–	–	–	–	–	–	–	–	–	–	–	–	–
4	–	–	–	–	–	–	–	–	–	–	–	–	–
5	–	–	–	–	15	22	16	4	–	–	–	–	5
6	–	–	8	35	64	81	64	37	21	6	–	–	26
7	4	17	70	136	168	197	178	149	116	75	32	8	96
8	67	97	177	257	275	312	295	278	240	185	122	71	198
9	143	183	263	356	384	405	390	376	331	273	215	164	290
10	204	244	361	424	436	475	460	444	407	336	265	222	357
11	258	290	403	431	450	504	505	466	423	376	296	244	387
12	277	297	430	454	450	514	515	485	428	386	315	270	402
13	278	299	418	424	450	504	505	475	414	370	284	241	389
14	215	243	356	390	407	460	467	442	378	307	228	195	341
15	146	173	278	342	350	381	402	372	293	217	138	114	267
16	54	94	171	231	248	287	303	266	171	102	32	17	165
17	–	13	59	117	136	172	179	139	52	4	–	–	73
18	–	–	–	8	34	60	60	28	–	–	–	–	16
19	–	–	–	–	–	–	–	–	–	–	–	–	–
20	–	–	–	–	–	–	–	–	–	–	–	–	–
21	–	–	–	–	–	–	–	–	–	–	–	–	–
22	–	–	–	–	–	–	–	–	–	–	–	–	–
23	–	–	–	–	–	–	–	–	–	–	–	–	–
Sum	1.646	1.951	2.993	3.608	3.867	4.373	4.341	3.962	3.275	2.638	1.927	1.544	

FIGURE 5.25 Hourly data per month for power from PV.

Hour \ Month	1	2	3	4	5	6	7	8	9	10	11	12	AVE
0	191	186	183	167	181	199	248	255	219	208	198	201	203
1	112	111	103	93	96	105	121	126	117	114	113	120	111
2	77	81	72	71	71	75	82	83	75	78	74	82	77
3	73	78	66	68	67	71	78	78	71	76	70	78	73
4	73	79	68	68	66	70	76	78	71	74	69	78	73
5	72	79	69	67	66	69	75	77	70	73	65	77	72
6	72	80	68	66	65	68	74	79	70	70	64	76	71
7	70	79	64	64	63	64	70	79	76	76	63	74	70
8	90	87	80	84	86	90	94	99	94	93	84	89	89
9	114	116	112	117	124	127	135	143	134	132	121	115	124
10	213	203	190	216	245	255	325	467	284	236	217	223	256
11	289	260	235	277	404	452	604	720	442	329	283	291	382
12	310	313	305	355	486	564	731	735	587	479	355	311	461
13	311	319	310	437	517	583	743	744	614	500	364	315	480
14	314	323	313	446	536	601	753	755	623	505	368	317	488
15	314	326	314	450	545	618	761	759	628	518	371	317	493
16	326	337	326	467	566	633	776	770	632	519	380	327	505
17	332	333	316	450	553	609	766	759	620	490	379	336	495
18	365	361	332	439	544	600	757	760	602	474	413	378	502
19	371	373	364	435	514	568	751	753	571	477	416	382	498
20	360	353	345	420	445	490	684	731	549	461	403	369	467
21	309	301	300	348	369	411	531	555	433	381	340	316	383
22	264	258	259	279	294	315	414	431	342	310	293	276	311
23	203	198	191	225	237	252	330	334	258	234	223	216	242
Sum	5.224	5.234	4.988	6.109	7.140	7.888	9.977	10.369	8.180	6.907	5.727	5.362	

Load [kW]

FIGURE 5.26 Hourly data per month for power to load.

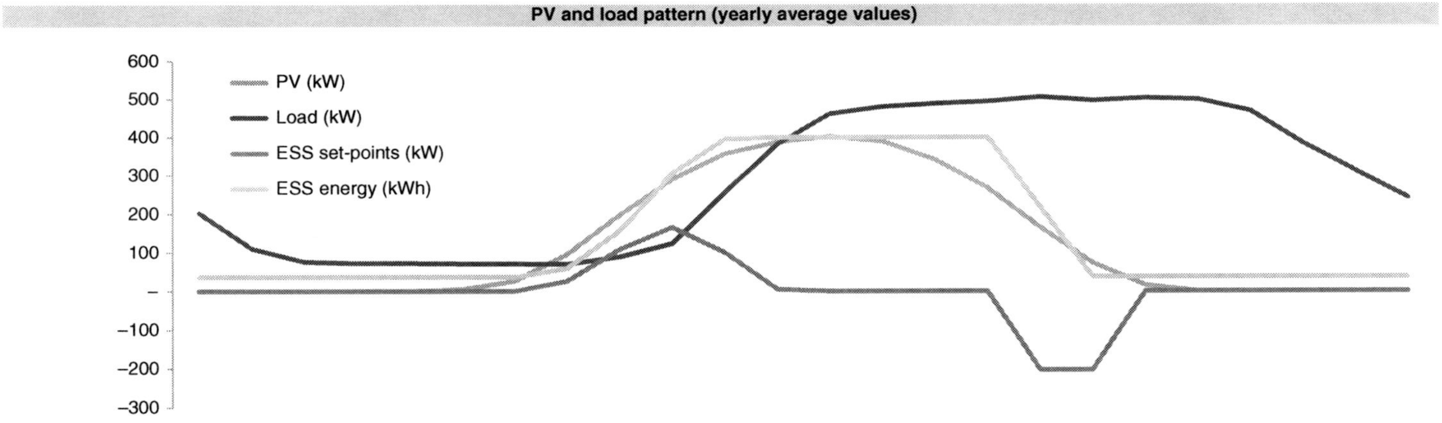

PV and load pattern (yearly average values)

Legend: PV (kW) · Load (kW) · ESS set-points (kW) · ESS energy (kWh)

Y-axis: 600, 500, 400, 300, 200, 100, –, –100, –200, –300

Hours	0	1	2	3	4	5	6	7	8	9	10	11	12	13	14	15	16	17	18	19	20	21	22	23
PV (kW)	–	–	–	–	–	5	26	96	198	290	357	387	402	389	341	267	165	73	16	–	–	–	–	–
Load (kW)	203	111	77	73	73	72	71	70	89	124	256	382	461	480	488	493	505	495	502	498	467	383	311	242
Delta (kW)	–203	–111	–77	–73	–73	–67	–45	26	109	166	100	5	–59	–91	–147	–226	–340	–422	–486	–498	–467	–383	–311	–242
Power to ESS (kW)	–	–	–	–	–	–	–	26	109	166	100	5	–	–	–	–	–	–	–	–	–	–	–	–
Power from ESS (kW) - 1	–	–	–	–	–	–	–	–	–	–	–	–	–	–	–	–	380	472	–	–	–	–	–	–
Check for C-rate limits	–	–	–	–	–	–	–	–	–	–	–	–	–	–	–	–	203	203	–	–	–	–	–	–
Check for available capacity	–	–	–	–	–	–	–	–	–	–	–	–	–	–	–	–	203	203	–0	–	–	–	–	–
ESS set-points (kW)	–	–	–	–	–	–	–	26	109	166	100	5	–	–	–	–	–203	–203	0	–	–	–	–	–
Energy to/from ESS (kWh)	–	–	–	–	–	–	–	23	97	149	90	4	–	–	–	–	–182	–182	0	–	–	–	–	–
ESS energy (kWh)	36	36	36	36	36	36	36	59	157	305	395	400	400	400	400	400	218	36	36	36	36	36	36	36
Command	G	G	G	G	G	G	G	C	C	C	C	C	G	G	G	G	D	D	G	G	G	G	G	G
Check	OK	OK	OK	OK	OK	OK	OK	OK	OK	OK	OK	OK	OK	OK	OK	OK	OK	OK	OK	OK	OK	OK	OK	OK

ESS sizing   400 kWh
PCS power   223 kW
Discharged energy 363 kWh
Storage specs

FIGURE 5.27 Simplified sizing for yearly average load and PV values.

PV and load pattern (monthly highest)

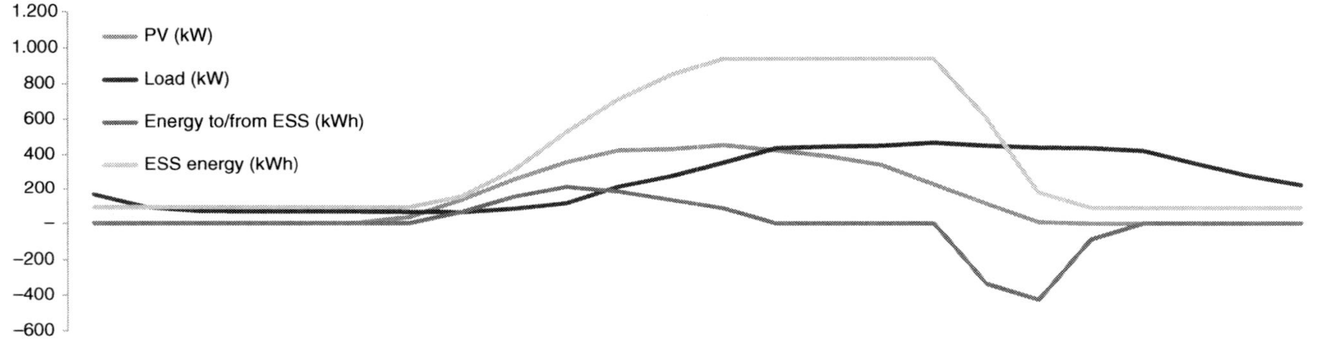

Hours	0	1	2	3	4	5	6	7	8	9	10	11	12	13	14	15	16	17	18	19	20	21	22	23
PV (kW)	-	-	-	-	-	3	35	136	257	356	424	431	454	424	390	342	231	117	8	-	-	-	-	-
Load (kW)	167	93	71	68	68	67	66	64	84	117	216	277	355	437	446	450	467	450	439	435	420	348	279	225
Delta (kW)	-167	-93	-71	-68	-68	-65	-31	72	173	239	208	154	99	-13	-56	-108	-236	-332	-431	-435	-420	-348	-279	-225
Power to ESS (kW)	-	-	-	-	-	-	-	72	173	239	208	154	99	-	-	-	-	-	-	-	-	-	-	-
Power from ESS (kW) - 1	-	-	-	-	-	-	-	-	-	-	-	-	-	-	-	-	-	371	482	486	-	-	-	-
check for C-rate limits	-	-	-	-	-	-	-	-	-	-	-	-	-	-	-	-	-	371	473	473	-	-	-	-
check for available capacity	-	-	-	-	-	-	-	-	-	-	-	-	-	-	-	-	-	371	473	101	-	-	-	-
ESS set-points (kW)	-	-	-	-	-	-	-	72	173	239	208	154	99	-	-	-	-	-371	-473	-101	-	-	-	-
Energy to/from ESS (kWh)	-	-	-	-	-	-	-	64	155	214	186	137	89	-	-	-	-	-332	-423	-90	-	-	-	-
ESS energy (kWh)	93	93	93	93	93	93	93	157	312	526	712	850	938	938	938	938	938	606	183	93	93	93	93	93
Command	G	G	G	G	G	G	G	C	C	C	C	C	C	G	G	G	G	D	D	D	G	G	G	G
Check	OK	OK	OK	OK	OK	OK	OK	OK	OK	OK	OK	OK	OK	OK	OK	OK	OK	OK	OK	OK	OK	OK	OK	OK

ESS sizing   930 kWh
PCS power   520 kW
Discharged energy 845 kWh
Storage specs
  ESS C-rate   2
  SoC min   10%
  RT efficiency   80%

FIGURE 5.28   Simplified sizing for monthly highest load and PV values.

Optimization entails deciding when the ESS should charge or discharge energy depending on the best combination of electricity prices, load needs, with a time view that must take into consideration forecasts at least within the 2- to 4-h period. A possible objective function for the optimization problem can be described by:

> maximize the revenues from peak-shaving and time-of-use
> subject to:
> > if power draw threshold is surpassed, discharge the battery;
> > if electricity prices are high and the load is sufficient, discharge the battery;
> > if the electricity prices are low, charge the battery from the grid;
> > maintain the ESU SOC within the limits;
> > do not increase power draw maximum threshold when charging;
> > . . .

The dynamics of the ESU are embedded in the optimization algorithm to take into account the real behavior of the system. The results for an ESS with power of 1 MW, energy of 1.5 MWh, a round-trip efficiency of 80%, a minimum SOC of 20%, a maximum SOC of 90%, a total charging time of 6 h, a discharge time of 2.5 h, and an industrial load with a maximum draw of 3.9 MW, are shown in Fig. 5.29, where from top to bottom are: the industrial load, the SP for the ESS, the variation of the SOC during charge and discharge, and the electricity price trend.

It is possible to notice that, when electricity prices are sufficiently high, the battery is discharging by a power SP depending on how much power must be supplied to the load, and is recharged when electricity prices are low. This optimization, running on data that are forecasts of actual values, have a higher uncertainty than the one on energy prices alone, due to the additional impact of the uncertainty of the load variations.

The optimization algorithm runs with a time granularity that depends on the system manager decision and system dynamics; where load variations or price volatility are high, the optimal SPs can be computed every 5–10 min. The SPs are used as a day-ahead planning tool and as an intraday tool, if electricity prices are timely updated by the provider of market prices.

5.10.4 Renewable Energy Injection Profiling Using Hydrogen Storage

A REN energy system coupled with a hydrogen ESS for injection profiling is shown in Fig. 5.30 [13].

The REN source is solar radiation converted to electric energy by the PV power plant. According to the control logic of the power plant EMS, the electric energy is converted into chemical energy with the electrolyzer (EL), and power is supplied to the load either directly by the PV field or by the fuel cell (FC). All devices are indirectly connected to the central bus-bar by means of buck and boost converters, temperature and pump controllers (COP) or through an inverter that converts DC to AC electric current.

The REN power plant is a large-scale, ground-based, fixed solar PV plant of 1.6 MW$_p$ comprising of 6880 modules with 235 W$_p$ nominal power and conversion efficiency of 14.3%. The overall installed electrolysis subsystem comprises of four electrolyzing units for a total installed power of 1.32 MVA and a total hydrogen flow of 3 mol/s. The fuel cell is a PEM with a net power output of 1.5 MW and a hydrogen consumption of 12.31 mol/s at a nominal efficiency of 40%. The electrolyzer and fuel cell system data are reported in Table 5.4, where $\eta_{F,el}$ is the Faraday efficiency of the electrolyzing system and η_{boost} and η_{buck} the conversion efficiency of the boost and buck converters.

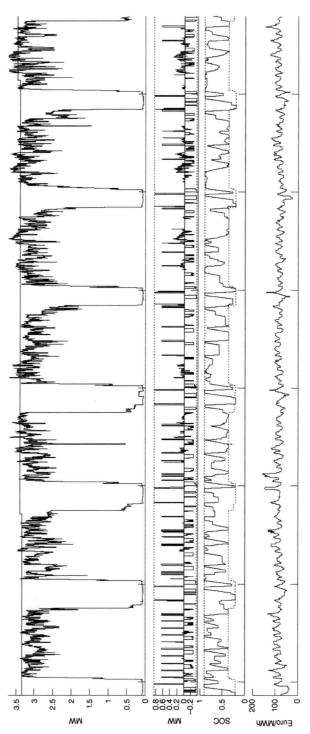

FIGURE 5.29 Example of optimal use of ESS for peak-shaving and time-of-use applications, depending on load and electricity prices.

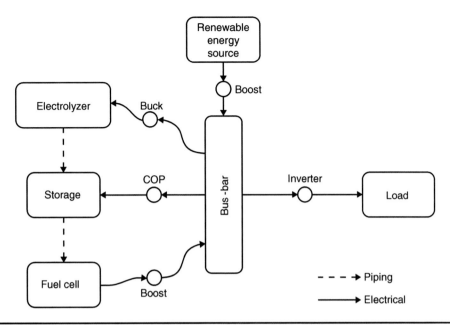

FIGURE 5.30   System schematic for REN profiling with hydrogen storage [13].

The conversion from DC to AC is performed by means of three inverters with nominal power 500 kW and maximum efficiency η_{max} of 98.5%.

The energy storage system is used to inject a known energy profile in the grid. The control logic uses day-ahead forecast energy data as the SP for the power to be injected in the grid. The power output is an on-off flat profile starting at 7 a.m. and ending at 5 p.m. Figure 5.31 outlines the profiles of the power converted by the PV system and the SP power injected in the grid.

When the power from the PV field is higher than the power SP, the electrolyzers receive the difference in power between the PV field power and the power to be injected in the grid. The electrolyzers then convert such energy from electrical to chemical. The

System	Type	Parameter	Value	Parameter	Value
Electrolyzer	Alkaline	$\eta_{F,el}$	80%	$U_{el,0}$	2.0268 V
		$C_{1,el}$ ($°C^{-1}$)	−0.0161	$I_{el,0}$	0.0122 A
		$C_{2,el}$	0.5011	R_{el}	0 Ω°C
		N_{cells}	360 per unit	Units	4
Fuel cell	PEM	*Round-trip efficiency*	40%	$U_{fc,0}$	2.0362 V
		$C_{1,fc}$ ($°C^{-1}$)	−0.0008	$I_{fc,0}$	0.5399 A
		$C_{2,fc}$	−0.0963	R_{fc}	0 Ω°C
		N_{fc}	1900		

TABLE 5.4   Electrolysis and Fuel Cell System Data [13]

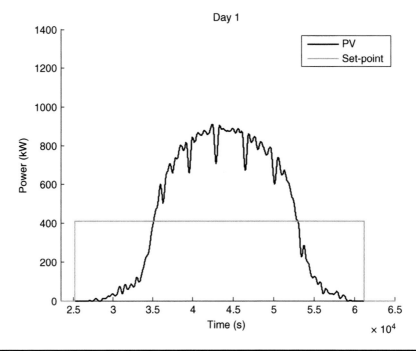

FIGURE 5.31 PV power and, in gray, the power injection SP for 1 day.

power injected in the grid is directly provided by the PV field. If the power from the PV field is lower than the SP, the fuel cells provide the difference between the SP and the PV field (see Table 5.5).

Equation (5.20) is used to balance the electrical currents in and out of the bus-bar:

$$I_{pv \to bus} - I_{bus \to el} + I_{fc \to bus} - I_{bus \to load} = 0 \qquad (5.20)$$

where $I_{pv \to bus}$ is the current flowing from the PV plant to the system bus–bar, $I_{bus \to el}$ the current from the bus-bar to the electrolyzer, $I_{fc \to bus}$ the current from the fuel cell to the bus–bar, and $I_{bus \to load}$ the current from the bus-bar supplied to the load. If $I_{pv \to bus} < I_{bus \to load}$ (the current $I_{bus \to load}$ is supplied to the load by a DC–AC converter with efficiency η_{inv}), the PV field would not convert enough energy to maintain the programmed

Condition	System	System status	Power to system
PV power > SP	Electrolyzer	ON	PV–SP
	Fuel cell	OFF	0
	Load	Supplied by power from PV only	SP
PV power <= SP	Electrolyzer	OFF	0
	Fuel cell	ON	SP–PV
	Load	Supplied by power from FC and PV	SP

TABLE 5.5 Outline of the Control Logic [13]

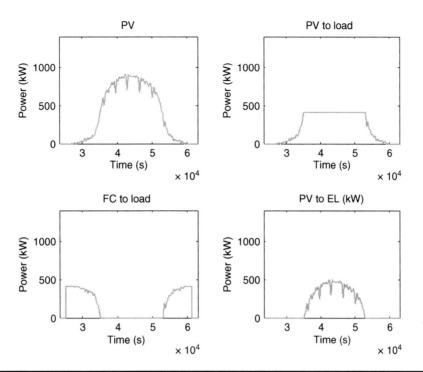

FIGURE 5.32 Daily trends for main subsystems.

injection curve and the fuel cell compensates by releasing $I_{fc \to bus}$ (the current $I_{fc \to bus}$ is supplied to the bus-bar by a boost-converter with efficiency η_{boost}); conversely, when $I_{pv \to bus} > I_{bus \to load}$, the surplus amount of energy from the PV plant is channeled to the electrolyzer to be stored as hydrogen (the current $I_{bus \to el}$ is supplied to the electrolyzer by a buck-converter with efficiency η_{buck}). The hydrogen storage systems acts as a filter that smooths out all the fluctuations that would be injected in the grid by the PV plant. This comes at the costs of the energy losses caused by the conversion steps of the hydrogen loop (namely, the energy conversion path of the electrolyzer, storage, and fuel cell). The electric energy is injected in the grid by means of a transformer that raise from low voltage to medium voltage levels typical of the distribution lines in the country (typically, 15–20 kV).

The efficiency of the system can be computed according to:

$$\eta = \frac{\int_0^T (P_{pv \to load} + P_{fc \to load} - P_{comp})dt + LHV \cdot \Delta n_{tank}}{\int_0^T P_{pv}dt} \tag{5.21}$$

where $P_{pv \to load}$ and $P_{fc \to load}$ are the PV and FC power supplying the load, before inverter conversion, according to Table 5.5, P_{comp} is the power to the compressor, LHV is the lower heating value of diatomic hydrogen, Δn_{tank} is the difference in hydrogen moles contained in the tank between end and beginning of period, and P_{pv} is the DC power output from the solar field before the inverter conversion to AC power. Such value depends, apart from all conversion stages of energy and relevant subsystem efficiencies, by the quota of PV that is directly provided to the grid (see Fig. 5.32).

5.10.5 Efficient Design and Control with Genetic Algorithms

The application of a GA has been performed on a hybrid energy system comprising of a solar PV and a wind energy conversion system, a diesel generator with a battery set and an hydrogen loop to provide electrical energy to a load [14, 15]. The optimization problem consists in the search for the components and control strategies that minimize the cost of the system. The GA consists of two sub-GA: the main algorithm finds the number of components that minimizes system capital expenses, the second the optimal control strategy for every combination of the system design to minimize operational expenses of the system as a result of the output of the main algorithm.

The genotype of the main algorithm is a vector of 11 integers: number of PV panels in parallel; type of PV panel; number of wind turbines; type of wind turbine; type of hydro turbine; number of batteries in parallel; type of battery; type of AC generator; type of fuel cell; type of electrolyzer; type of inverter. Some key components (i.e., the charge regulator, the rectifier, and the hydrogen tank) are not part of the vector of the main algorithm since their optimal size is determined once the secondary algorithm provides the best control strategy.

The main algorithm uses a population of N_m vectors, the individuals, and a number of generations to be evaluated that is capped at $N_{genMainMax}$. For each individual, the system cost is evaluated, the individuals are ranked accordingly, and a fitness function is applied to the individuals according to their ranking. For the ith individual, the fitness function is:

$$\text{mainFit}_i = \frac{(N_m + 1) - i}{\sum_j [(N_m + 1) - j]}, \quad j = 1 \dots N_m \tag{5.22}$$

where j is the rank of the individual in the population, 1 being the best, or *fittest*, individual. The fitness function determines the probability of selecting an individual on which to apply the genotype mutation during the reproduction step of the GA.

The secondary algorithm is devised to find the control strategy that minimizes, for each configuration provided by the main algorithm, the operating cost function. Its genotype consists of 12 control variables: minimum power of the AC generator recommended by the manufacturer; minimum power of the fuel cell recommended by the manufacturer; minimum state of charge of the battery recommended by the manufacturer; AC generator critical power limit; batteries SOC set point for the AC generator; fuel cell critical power limit; batteries SOC set point for the fuel cell; set point for the amount of H_2 stored in the tank; power below which it is more economical to store energy in the batteries than in the H_2 tank; intersection point of the cost of supplying energy with the batteries and the cost of supplying energy with the fuel cell; and intersection point of the cost of supplying energy with the fuel cell and the cost of supplying energy with the AC generator.

Everything works the same as per the main algorithm, and similarly to the main algorithm, the fitness function of the ith individual is given by:

$$\text{secFit}_i = \frac{(N_s + 1) - i}{\sum_j [(N_s + 1) - j]}, \quad j = 1 \dots N_s \tag{5.23}$$

where j is the rank in the population (1 being the fittest individual).

Elitism is the rule common to both algorithms: best individuals are not lost from one generation to the next.

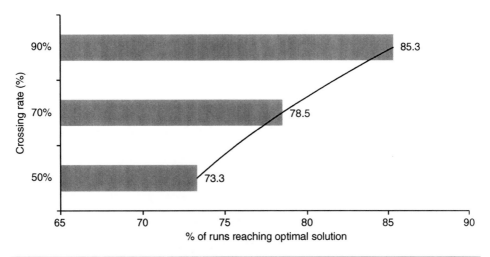

FIGURE 5.33 Percentage of projects reaching the global solution by different crossing rates (adapted from [15]).

Gene evolution is performed by means of either a "uniform" or a "nonuniform" mutation. The gene to be modified is always chosen with the same probability distribution per each gene in the genotype of the individual. When the gene has been selected, the mutation can occur as per the following. A *uniform mutation* is achieved by applying to a random gene a value obtained by using a uniform distribution of probability in the interval $[LL, HL]$ where LL and HL, respectively, are the lower and the higher limit for the same gene. A *nonuniform mutation* occurs by changing a randomly selected gene g by using the following:

$$g + \Delta(N_{\text{genMain}}, HL - g) \text{ if a random number binary number is 0}$$
$$g + \Delta(N_{\text{genMain}}, g - LL) \text{ if a random number binary number is 1}$$
(5.24)

where N_{genMain} is the number of the generation, HL and LL are the higher limit and the lower limit of the variable represented in the gene g. The function $\Delta(N_{\text{genMain}}, y)$ produces a value in the interval $[0, y]$ so that the probability of $\Delta(N_{\text{genMain}}, y)$ approaching 0 increases as N_{genMain} increases. This operator initially searches solutions globally, then locally in the last generations in order to increase the likelihood of generating individuals with genotypes closer to its successor. It is written as:

$$\Delta(N_{\text{genMain}}, y) = y \left(1 - r^{\left(1 - \frac{N_{\text{genMain}}}{N_{\text{genMainMax}}} \right)^b} \right)$$
(5.25)

where r is a random number in the range $[0, 1]$, $N_{\text{genMainMax}}$ is the maximum generation number, and b is a parameter determining the degree of dependency upon the generation number to be chosen by the designer. In this example, b has been chosen as being equal to 0.5.

Figure 5.33 gives the percentage of projects that reach the optimal solution by different crossing rates.

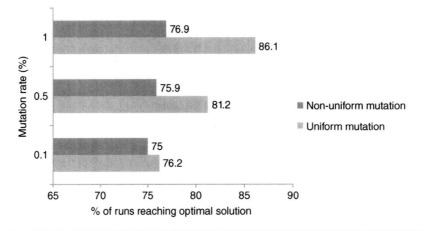

FIGURE 5.34 Optimal solution with varying mutation rates and mutation typologies (adapted from [15]).

Figure 5.34 gives the percentage of projects that reach the optimal solution by changing the mutation rates and type of mutation.

To facilitate reaching a close-to-optimal solution, the algorithm can be designed and run with the following settings:

- number of generations higher than 15
- population size higher than 0.003% of the number of total combinations
- crossing rate 90%
- mutation rate 1%
- uniform mutation

5.10.6 Sizing Optimization of a Stand-Alone Lighting System with Genetic Algorithms

A practical application of a combination of a GA with a simplex algorithm is described in [16] where system parameter and cost optimization is performed for a stand-alone photovoltaic hydrogen hybrid system supplying a street lighting load.

Since power production depends on variable weather conditions, deterministic algorithms can be difficult to be modeled; on the contrary, a heuristic global search like GA is easily defined to look for a set of individuals whose genotype will become the initial data set that will be further refined by the SA. The genotype of the individuals is set to be the vector $\{P_{pv}, Q_h, P_{fc}, \beta, SOC_{min}, SOC_{max}\}$ where:

- P_{pv} : photovoltaic plant power
- Q_h: battery capacity
- P_{fc}: fuel cell power
- β: tilt angle for PV modules
- SOC_{min}, SOC_{max}: minimum and maximum SOC

The function to be optimized is an economic cost function. System parameters are to be determined by evaluation of penalties related to excess or shortage of energy. GA is designed as:

- 15 individuals per generation
- 100 generations
- Roulette selection scheme
- 47% mutation probability

The fitness is defined as a percentage so that, if the total cost tends to zero, fitness tends to 100%. GA is employed in combination with a traditional deterministic algorithm, the strength of soft computing techniques in finding solution through heuristics joining forces with a more traditional algorithm to increase final solution accuracy.

By comparing the results of a GA alone with a SA + GA, the end costs are minimized by a further 32%, with significant differences in the configuration of the system between the solution given by the GA and the following refinement with the SA. The PV power plant from the initial 95 kW_p is refined by the SA to an increased 148 kW_p, while the fuel cell system power decreases to 128 kW_p from 282.6 kW_p. Finally, overall computing time is decreased by the synergy between a combined heuristic and deterministic approach.

References

1. www.modbus.org.
2. tc57.iec.ch/index-tc57.html.
3. Webster J. (Ed.) (2007). Wiley Encyclopedia of Electrical and Electronics Engineering. John Wiley & Sons, Inc: USA.
4. Rashid M. H. (Ed.) (2011). Power Electronics Handbook–Devices, Circuits, Applications. Elsevier.
5. Meegahapola L., Littler T., Perera S. (2013). Capability curve based enhanced reactive power control strategy for stability enhancement and network voltage management. International Journal of Electrical Power and Energy Systems, 52(1):96–106.
6. (2008). Oxford English Dictionary. Oxford University Press: UK.
7. Cavazzuti M. (2013). Optimization Methods, From Theory to Design, Scientific and Technological Aspects in Mechanics. Springer-Verlag: Berlin, Heidelberg.
8. Author: Marcuswikipedian, File:Pareto Efficient Frontier for the Markowitz Portfolio selection problem..png, Site: commons.wikimedia.org, Date: 29th of February 2012, Licence: Creative Commons 3.0.
9. Nocedal J., Wright S. J. (2006). Numerical Optimization. Springer-Verlag: Berlin, Heidelberg.
10. Zini G., Pedrazzi S., Tartarini P. (2011). Use of soft computing techniques in renewable energy hydrogen hybrid systems. In: Soft Computing in Renewable Energy Systems, Gopalakrishnan K. et al. (Ed.), Springer, pp. 37–64.
11. Schmitt Lothar M. (2001). Theory of generic algorithms. Theoretical Computer Science 259(1–2):1–61.
12. Eiben A. E., Smith J. E. (2007). Introduction to Evolutionary Computing (Natural computing series). Springer: Berlin.

13. Zini G., Dalla Rosa A. (2014). Hydrogen systems for large-scale photovoltaic plants: Simulation with forecast and real production data. International Journal of Hydrogen Energy 39:107–118.

14. Dufo-Lòpez R., Bernal-Agustìn J. L., Contreras J. (2007). Optimization of control strategies for stand-alone renewable energy systems with hydrogen storage. Renewable Energy 32:1102–1126.

15. Bernal-Agustín J. L., Dufo-Lòpez R. (2009). Efficient design of hybrid renewable energy systems using evolutionary algorithms. Energy Conversion and Management 50:479–489.

16. Lagorse J., Paire D., Miraoui A. (2009). Sizing optimization of a stand-alone street lighting system powered by a hybrid system using fuel cell, PV and battery. Renewable Energy 34:683–691.

PART II

Making It Happen: Economy and Finance for Energy Storage

CHAPTER 6

Financial, Economic, and Risk Analysis

Summary.   This chapter provides the reader with a set of concepts and tools that are used to evaluate the financial and economical viability of energy projects. From the construction of a financial model, to its analysis and to the risk assessment of the project, everything that is introduced here is actually used in real life and have been proven valid over and over in the deployment of many energy plants currently in operations.

6.1   Introduction

When considering the development of an energy storage system, its full economic impact must be assessed to make sure that the chosen storage technology provides the benefits requested by the decision makers of the organization. The business model that is constructed for the project is needed by the organization to reach a decision on whether allocating capital to the project, and by the financing institutions to understand if and how to invest in the opportunity. For the same reason, a good analysis must also entail the evaluation of risk that the projects embed, and its disclosure to managers in order to understand and address potential issues as the project is developed into an operating asset.

Therefore, knowledge of the right tools is crucial to enable the organizations to make the shift from a mere blueprint design to the make-it-happen, the final realization of the project, and assure its impact on the real world.

This chapter is written to provide readers with the tools that are used by the professionals in the industry when confronted with capital allocation problems in energy storage projects. All the techniques and concepts come from real-world usage and have proven to be indispensable to deploy a viable and sustainable energy project, and engineering projects at large.

Of course, the information provided here in not intended as a substitute of specialized finance and economics books, to which the interested reader is referred to in the references at the end of this chapter.

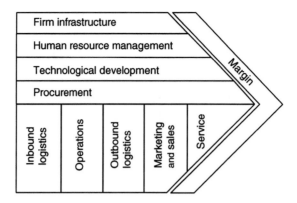

FIGURE 6.1   The value chain of a firm.

6.2 Concepts of Corporate Finance

Corporate finance is best described as the discipline of economics and management that has the objective of maximizing the value of the firm [1, 2]. Corporate finance deals with the management of the money that flows in and out the firm rather than with manufacturing, operations, marketing, health and safety, logistics, but has a direct, prominent impact on all the value chain of the firm itself (Fig. 6.1). No firm can live without some form of proper corporate financial management.

Corporate finance provides the tools to assist the firm development in many facets, but its basic principles [3] can be singled out as:

- the choice of the financing sources, and their combination, that can minimize the *hurdle rate*[1] and that are consistent with the assets of the firm (the *financing principle*) during all the different development stages and life cycle of the organization, from startup to maturity.

- the decision whether investing in projects that yield a return greater than the hurdle rate (the *investment principle*), or

- in case there are not enough investments available to the firm that earn the hurdle rate, return the cash to the owners of the firm (the *dividend principle*).

The money that is provided to the firm come as *Equity*, the money that is provided to the firm by the owners of the firm, and *Debt*, the money that is received by third parties, with no property on the firm.

Providing money to the firm does not come without costs. Equity is rewarded by the *cost of equity*, which is the return that the firms owes to its owners to reward them of the risk of investing in the firm. Money lenders, like banks, financial institutions, venture capital, or public organizations are rewarded by the *cost of debt*, which materializes in an interest rate that is requested by lenders on the total amount of the money lent.

---

[1] The *hurdle rate* is the minimum return on the investment that is deemed acceptable by the managers or the investors.

Static analysis	
	Annual reports
	Reclassification
	Ratio analysis
Dynamic analysis	
	Cash flow statement
Prospective analysis	
	Cash budget
	Financial plan
	Working capital management

TABLE 6.1   Outline of Corporate Finance Analysis

Corporate finance entails the performance of certain types of analysis to help the management team in the evaluation of the firm's situation and provide them with relevant information for decision making.

Static analysis is the analysis of the performance of the firm and how well the firm has been conducted by its managers over a period of time with respect to other firms, usually the competitors, and with respect to the firm itself over the years. The tools that are used to perform static analysis are the *financial statements* which are prepared annually. Such reports are reclassified to perform in-depth analysis according to different goals of the analysis itself. From reclassification, the *ratio analysis* is applied to extract knowledge from the wealth of data embedded in the annual reports.

Dynamic analysis seeks to understand and optimize the cash generated or absorbed by the firm: the *cash-flow statement* is the main tool adopted in this step.

Prospective analysis looks at the future of the firm, namely where is the organization going. Short term (1 or 2 years) is analyzed by means of the *cash budget*, while the medium to long-term period (3–5 years) concerns the design of a financial plan. Both short and long term are also entailing the analysis and management of the working capital needs of the firm.

Table 6.1 provides an outline of the corporate finance analysis and related tools.

From the tools and techniques of corporate finance, the management derives many decisions about the future of the firm, in order to maximize the value of the organization by the coordination of all the firm's value chain functions.

6.3   Financial Statements, Reclassification, and Ratio Analysis

Annual reporting is compulsory for many organizations and is needed as a mean to communicate the financial situation to the external interested parties. Such communications are requested by law, and reports are prepared according to standards that can vary between countries. Reporting is not only needed to comply with regulations, but is also a fundamental tool for management to understand the situation and create action after data analysis. This is why there can exist two different reporting documents; the external, official, standardized reporting, and an internal reporting based on ad hoc structures, used internally for data analysis.

All documents are based on data collection and organization from the accounting department or external consultants. Depending on a firm's dimension, annual reporting must be assessed and certified by third-party, external advisory companies.

Annual reports comprise of the set of financial statements which are: the balance sheet, the income statement, and their explanatory notes. The financial statements are drafted in compliance with the *Generally Accepted Accounting Principles* (GAAP) or *International Accounting Standards* (IAS), depending on the country. Alongside official structuring of the financial statements, also referred to as "external" financial statements, the financial statements can be reorganized in "internal" reports to help management evaluate the business situation of the firm according to *reclassification* principles that have become standard management practice.

The *explanatory notes* are documents that provide information on how the balance sheet and income statement have been constructed, or clarifications on particular accounting calculations, how reserves have been generated or disbursed, changes in tangible or intangible assets, and all pieces of information that are deemed necessary to make the analysis of the financial statements the clearest possible by all interested readers.

6.3.1 Balance Sheet

The *balance sheet* (B/S) is the snapshot of the firm situation over one year of operations. It is divided into two sections, that match together when all items in each section are summed together. One section represents the *assets*, which are all the means that a firm has in its availability for use. The other section concerns the *liabilities*, namely what the organization has received from its owners or third parties and that the organization owes to such parties comprising equity, where the following equation holds true:

$$\text{Assets} = \text{Equity} + \text{Liabilities} \tag{6.1}$$

The sum of all assets is also referred to as the *total uses* of the firm, whereas the sum of all liabilities and equity is the *total sources* of the firm.

A first distinction is made on the time frame of the assets and liabilities: short or long term. Short-term assets are all the assets that a company can confidently sell and convert into cash in a 1-year period. Short-term liabilities are all liabilities that are due to creditors within a 1-year period. Long-term assets and liabilities are therefore all other items that cannot materialize as cash in a 1-year period.

Within the short-term assets class:

- *Cash and cash equivalents* are cash in bank account, and highly liquid and easy to sell financial instruments with insignificant risk of change in value, like government bonds or commercial papers from solid and high rating firms. Cash equivalents are also referred to as *market securities*.

- *Inventory* is the raw materials, half-finished goods that are in process within the operations of the firm.

- *Accounts receivable* are invoices to customers for the sale of goods or services, and represent legally enforceable claims.

- *Prepaid expenses* are payments made by the firm in advance to receiving goods or services.

- *Accrued revenues* are revenues not yet received by already matured and that will be received when due in the near future (i.e., a monthly rent).
- *Notes receivable* are legally enforceable formal claims issued as evidence of a debt, like promissory notes, that lead to payment reception within a short period of time.

Within the long-term assets class:

- *Tangible assets* like plant, equipment, machinery etc
- *Intangible assets* like patents or goodwill, the latter representing a measure of the value of the firm which is not directly linked to its tangible assets but, for instance, to brand identity or market recognition
- *Financial assets* are formal instruments like bank deposits, bonds, stocks, or contractual rights to receive cash when exchanged in the financial markets

Long-term assets are taken net of *depreciation*, which is the loss in value of tangible assets, and *amortization*, the loss in value of intangible assets over their life cycle (depreciation and amortization often abbreviated as D&A). For instance, soon after purchasing, plants, equipment, but also patents, reduce their value due to aging from utilization or obsolescence. This reduction in value is recognized by the GAAP and is treated as a cost, although D&A is not a real cash outflow but merely an accounting entry.

Within the short-term liabilities class:

- *Current portion of loans/leases*
- *Overdrafts* in bank accounts
- *Accounts payable* are invoices made by suppliers for the purchase of goods or services, and represent legally enforceable claims
- *Unearned revenues* are revenues already received by the firm in advance to providing goods or services
- *Accrued expenses* are expenses not yet paid by already matured and that will be paid when due in the near future (i.e., a monthly rent)
- *Notes payable* are legally enforceable formal claims received as evidence of a debt, like a promissory notes, that lead to payment within a short period of time
- *Income tax payables*

Within the long-term liabilities class:

- *Loans/leases*
- *Bonds*
- *Shareholder loan* is a form of debt provided by the shareholders of the firm
- *Allowances, funds. . .* are liabilities that are maintained by the firm to cover, for example, staff severance funds or indemnities

Within the equity class:

- *Shareholder capital* is the capital endowed by the owners to the firm.
- *Retained earnings* are the sum of the profits and losses over time less the dividends paid to shareholders.
- *Reserves* are the apportionment of retained earnings that must be kept from being distributed to shareholders and needed to cater for special provisions or to comply with legislation. Depending on their use, reserves can be requested to be set apart as a percentage of the profits.

B/S is normally reclassified to help in the analysis and decision making (Fig. 6.2). For example, reclassification according to *liquidity-based* criteria organizes all accounting entries according to their maturity, and is used to help with the analysis of liquidity of the firm and its financial performance. Reclassification according to *activity-based* criteria is used mostly for operating performance analysis, by separating accounting entries depending on their core or noncore contribution to the firm's operations.

6.3.2 Income Statement

The *income statement* (I/S), also called *profit and loss* (P&L), is a report of the operations over a period of time, showing the revenues and the expenses incurred during a period of normally one year. While the B/S is showing a specific moment in time of the life of the firm, the I/S is a flow of money over time. The typical structure of the I/S is shown in Table 6.2.

The *earnings before interest, taxes, depreciation and amortization* (EBITDA), also known as *gross margin*, is what remains after costs have been taken out of revenues. Similarly, the *earnings before interest and taxes* (EBIT), or *operating margin* is what is left from EBITDA after depreciation and amortization have been taken out. The *net profit/loss* is what results as the bottom line of the I/S. While GAAP and IAS provide the official guidelines to draft the official I/S for any firm, more interesting to management are the internal I/S which are obtained through reclassification in the same way done for the B/S (Fig. 6.3).

Two ways to reclassify the I/S are normally employed. One is the *value-added* I/S, which focuses on the production activity of the firm, and is helpful in studying purchasing, production, marketing strategies. A second reclassification is performed according to *activity-related* criteria that classify data as core, commercial, marketing, financial, and extraordinary areas of profit or loss.

In the I/S, the *work in process* (WIP) is the half-finished goods or products that the firm manufactures.

The *cost of goods sold* (COGS) is the money value of the difference between initial and ending inventories, plus purchases, D&A, energy, lease payments, and similar disbursements that are functional to the core activities leading to the production of goods.

The *general and administrative* (G&A) expenses are the disbursements for wages, administrative, marketing, commercial, also known as "overhead" costs, all the costs that are functional to the core business of the firm.

6.3.3 Ratio Analysis

B/S and I/S are important sources of information for the management of a firm. A set of metrics has been developed over time to help in the analysis of the performance of

Balance sheet, liquidity based

Assets	Liabilities
Short term	**Short term**
Cash	Current portion of loans/leases
Inventory	Overdrafts
Accounts receivables	Accounts payables
Prepaid expenses	Unearned revenues
Accrued revenues	Accrued expenses
Notes receivable	Notes payable
Market securities	Income tax payables
Long term*	**Long term**
Plant	Loans/leases
Equipment	Bonds
Financial	Shareholder loans
other tangible assets	Allowances, funds**
Patents, goodwill	**Equity**
other intangible assets	Shareholder capital
	Retained earnings
	Reserves

Balance sheet, activity based

Assets	Liabilities
Core	**Core**
Cash	Income tax payable
Inventory	Allowances, funds**
Accounts receivables	Accounts payables
Prepaid expenses	Unearned revenues
Accrued revenues	Accrued expenses
Notes receivable	Notes payable
Noncore	**Noncore**
Plant	Current portion of loans/leases
Equipment	Accounts payables
Financial	Overdrafts
Patents, goodwill	Unearned revenues
other intangible assets	Accrued expenses
Not related to op. activities:	Notes payable
Cash	Bonds
Prepaid expenses	Shareholder loans
Accrued revenues	Loans/leases
Notes receivable	**Equity**
Market securities	Shareholder capital
	Retained earnings
	Reserves

* Net of depreciation and amortization
** That is, staff severance funds, indemnities....

FIGURE 6.2 Balance sheet, reclassified according to liquidity and activity based.

Revenues
– Costs
EBITDA (gross margin)
– Depreciation/amortization
EBIT (operating margin)
– Taxes
– Interests
Net profit (loss)

TABLE 6.2 Basic Structure of the Income Statement

the firm. Since these metrics are mostly ratios between numbers taken from the financial statements, *ratio analysis* is the activity performed on the financial statements to obtain information useful for decision-making.

Normally, the ratios are defined and calculated as a mean to compare the performance of the firm 1 year after the other, or as a comparison to other firms in a similar industry. This way, ratios are a very important source of information, whether for an internal comparative or an external competitive analysis.

Income statement, value added

+ Revenues
± Variation of inventories (finished goods)
Value of production/value of goods manufactured

–Net purchases (raw materials and WIP)
± Variation of inventory (raw materials and WIP)
–Other expenses
Value added

–G&A expenses
Gross operating margin (EBITDA)

–Depreciation and amortization
Operating margin (EBIT)

± (Financial revenues – expenses)
± (Extraordinary revenues – expenses)
± (Other revenues – expenses)
EBT

–Taxes
Net profit (loss)

Income statement, activity related

+ Revenues
–Cost of goods sold (COGS)
Industrial margin

–Commercial, marketing expenses
–G&A expenses

Operating margin (EBIT)

± (Financial revenues – expenses)
± (Extraordinary revenues – expenses)
± (Other revenues – expenses)
EBT

–Taxes
Net profit (loss)

COGS:
    + beginning inventory
    + purchases
    + depreciation
    + lease payments
    + energy
    + other expenses
    –beginning inventory

FIGURE 6.3 Income statement, reclassified as valued added or activity related.

Ratios are classified according to which goal they want to attain. Examples of ratios are given in the following list according to their area of intervention:

- Financial performance: return on equity, earning per share, price/earnings.
- Operating performance: profit margin, return on asset, asset turnover, revenue per employee, accounts receivable/payable turnover, plant/inventory turnover.
- Risk measurement: current ratio, quick ratio, debt to equity, interest coverage.

A detailed analysis of the ratios is outside the scope of this book, but the description and use of some of the most helpful ratios for energy project analysis is provided in Sec. 6.8.

6.4   Cash-Flow Statement

The *cash-flow statement* provides additional information to the financial statements discussed in Sec. 6.3. Cash-flow statement is the tool used to evaluate how much cash flows in or out of the organization from operating, investing or financing activities.

The *cash flow from operating activities* gives the cash generated, or absorbed, by variations in current assets like working capital, general operating expenses like wages to employees, taxes. If, for instance, more accounts receivable than accounts payable are received, the net cash is positive for the firm and represents a cash generation.

The *cash flow from investing activities* is the money freed or captured by investments or divestments of, for instance, real assets (land, building, and equipment), or investment securities.

The *cash flow from financing activities* represents the cash received or disbursed due to debt financing or debt repayment.

Figure 6.4 shows the construction of the cash-flow statement.

Cash-flow statement

EBITDA
– Taxes   Cash flow from operating
± Net working capital

+ Divestments   Cash flow from investing
– Investments

+ Financing
– Repayments   Cash flow from financing
+ Interests income
– Interest expenses

Net cash flow available to shareholders

– Distributions
± Other ancillary, extraordinary revenues/expenses
Net cash flow available to the firm

FIGURE 6.4 Cash-flow statement.

The final result is the information on how each area contributes to cash inflows or outflows. The comparison of the amounts recognized to each area, tells the firm where and how much cash enters or exits the firm.

Just to give a very simple example, if cash were only generated by the operating area, contribution to cash generation will be also needed from financing, otherwise, company growth could result to be hampered.

6.5 Cash Budget

A very important concept in the life of a firm is that having profits does not necessarily means that a firm is successful in what it's doing; cash is also extremely important because a firm can be profitable but still run out of cash. A firm receives profits in the future, but needs cash now; this time lag between the moment when a firm receives cash in the future (from profits), and its use of cash now in the present (to pay wages, raw material, components, services...) must be correctly managed by the firm. If the firm doesn't have money to pay for raw materials or components to produce its finished goods, the firm cannot manufacture and sell the products, and will not receive profits in the future. Furthermore, if a firm cannot pay wages, bank interests, or suppliers, someone will sooner or later claim that money and the firm become bankrupt.

A way a firm has to manage in and out cash flows correctly is the *cash budget*, the tool that estimates on a monthly basis all the cash being received from third parties (inflows) and being provided to third parties (outflows). The goal is to identify cash surpluses or deficits, to use surplus when possible or cover deficit in an effort to always make sure that monthly cash budget never becomes negative.

Data entry for the cash budget comes from the B/S and I/S, from operational budgets like production or sales budgets, and from other sources of information like new investments or divestments or new financing plans. All flows are actual forecasts of real values, meaning that cash budget must be considered a rolling exercise that must be updated periodically to maintain control over the cash situation.

To finance the cash deficit, normally cash in excess or short-term financing are used. If short-term financing is employed, the relative interest expenses must be included in the cash budget and considered as an additional cash outflow.

Typical cash inflows are given by:

- sales of goods
- accounts receivables reaching maturity
- operational divestments (sale of land, equipment, machinery, plants...)
- financial divestments (sale or issuance of bonds, shares...)
- equity increases
- credit lines from banks
- shareholders loans
- dividends, interests, royalties
- other inflows

Cash outflows can be:

- operational purchases
- operational payments (wages, insurances, utilities, rent...)
- accounts payable reaching maturity
- other G&A costs
- financial payments (interests, dividends, loan repayments...)
- taxes or payments to authorities
- other outflows

A simple example model for cash budget is given in Fig. 6.5.

6.6   Free Cash Flow

The *free cash flow* (FCF) is the cash that is available to the firm to be distributed to its shareholders according to the dividend principle as in Sec. 6.2, or to investments in operations, according to the investment principle.

The basic way to calculate the free cash flow is shown in Table 6.3.

When the management needs to understand the cash that can be available for distribution to shareholders, the *free cash flow to the equity* (FCFE) model is employed (see Fig. 6.6). The model focuses on financial data, dividing items in discretionary and nondiscretionary, and can provide insights on where to invest the excess cash or look for financing if FCFE is negative or not sufficient for internal and external growth strategies.

When the management needs to evaluate investments in operations, rather than distributing cash to shareholders, the *free cash-flow to the firm* (FCFF) is computed with the "operative cash-flow model" (see Fig. 6.6). The focus is on cash that is generated or absorbed by the firm, similar to the cash-flow statement in Sec. 6.4, and is capable of showing both operating and financial risk of the firm. Normally, the FCFF is used to decide upon the right mix of financing for diversification or internal growth.

The entries between the "free cash-flow line" and the "cash budget of the period" entail the item "new funding," which is the entry that is managed to find the right financing mix for the firm to avoid having negative cash budgets.

6.7   Project Valuation

Project valuation, or *capital budgeting*, aims at obtaining a quantitative measure of the returns of a project before its development, construction, and operation. Details on how to perform the valuation can be found in a number of specialized publications [4–6].

Running a thorough and accurate analysis is mandatory; a project often involves the utilization of large capital resources for a long period of time, with expenses that can be mostly paid upfront and returns that will only materialize in the future. Even in case of a project that is discontinued before its intended natural end of life, the amount of resources that have been used at the beginning, that is, debt, can remain stuck inside the

Cash budget

	Month 1	Month 2	Month 3	Month 4	Month 5	Month 6	Month 7	Month 8	Month 9	Month 10	Month 11	Month 12	Total
Beginning cash balance	€ 900.00	€ 299.00	€ 329.00	€ 305.00	€ 303.00	€ 392.00	€ 404.00	€ 374.00	€ 367.00	€ 356.00	€ 424.00	€ 374.00	€ 900.00
Cash from operations	€ 125.00	€ 200.00	€ 150.00	€ 160.00	€ 250.00	€ 160.00	€ 150.00	€ 187.00	€ 167.00	€ 246.00	€ 132.00	€ 184.00	€ 2,111.00
Total available cash	€ 1,025.00	€ 499.00	€ 479.00	€ 465.00	€ 553.00	€ 552.00	€ 554.00	€ 561.00	€ 534.00	€ 602.00	€ 556.00	€ 558.00	€ 3,011.00
Less:													
Capital expenditures	€ 670.00	€ 123.00	€ 113.00	€ 106.00	€ 102.00	€ 94.00	€ 114.00	€ 126.00	€ 133.00	€ 126.00	€ 119.00	€ 123.00	€ 1,949.00
Interest	€ 20.00	€ 26.00	€ 24.00	€ 25.00	€ 21.00	€ 22.00	€ 27.00	€ 29.00	€ 20.00	€ 27.00	€ 23.00	€ 29.00	€ 293.00
Dividends	€ 2.00	€ 4.00	€ 3.00	€ 5.00	€ 0.00	€ 2.00	€ 1.00	€ 2.00	€ 3.00	€ 2.00	€ 4.00	€ 5.00	€ 33.00
Debt repayment	€ 50.00	€ 20.00	€ 40.00	€ 32.00	€ 48.00	€ 39.00	€ 46.00	€ 41.00	€ 34.00	€ 30.00	€ 40.00	€ 42.00	€ 462.00
Other	€ 0.00	€ 0.00	€ 0.00	€ 0.00	€ 0.00	€ 0.00	€ 0.00	€ 0.00	€ 0.00	€ 2.00	€ 0.00	€ 0.00	€ 2.00
Total disbursements	€ 742.00	€ 173.00	€ 180.00	€ 168.00	€ 171.00	€ 157.00	€ 188.00	€ 198.00	€ 190.00	€ 187.00	€ 186.00	€ 199.00	€ 2,739.00
Cash balance (Deficit)	€ 283.00	€ 326.00	€ 299.00	€ 297.00	€ 382.00	€ 395.00	€ 366.00	€ 363.00	€ 344.00	€ 415.00	€ 370.00	€ 359.00	€ 272.00
Add:													
Short-term loans	€ 6.00	€ 3.00	€ 6.00	€ 6.00	€ 10.00	€ 9.00	€ 8.00	€ 4.00	€ 7.00	€ 9.00	€ 4.00	€ 5.00	€ 77.00
Long-term loans	€ 10.00	€ 0.00	€ 0.00	€ 0.00	€ 0.00	€ 0.00	€ 0.00	€ 0.00	€ 5.00	€ 0.00	€ 0.00	€ 0.00	€ 15.00
Capital stock issues	€ 0.00	€ 0.00	€ 0.00	€ 0.00	€ 0.00	€ 0.00	€ 0.00	€ 0.00	€ 0.00	€ 0.00	€ 0.00	€ 0.00	
Total additions	€ 16.00	€ 3.00	€ 6.00	€ 6.00	€ 10.00	€ 9.00	€ 8.00	€ 4.00	€ 12.00	€ 9.00	€ 4.00	€ 5.00	€ 92.00
Ending cash balance	€ 299.00	€ 329.00	€ 305.00	€ 303.00	€ 392.00	€ 404.00	€ 374.00	€ 367.00	€ 356.00	€ 424.00	€ 374.00	€ 364.00	€ 364.00

FIGURE 6.5 An example of a cash budget.

EBIT
+ Depreciation/amortization
− Taxes
± Net working capital
− Capital expenses
Free cash flow

TABLE 6.3 Basic Free Cash-Flow Calculation

project and create dangerous liabilities on the investing company. Risks must therefore be well understood and correctly weighed against the potential benefits of the construction and operation of the energy project.

Project valuation entails the use of assumptions and forecasts that are, by their very nature, only assumptions and forecasts. As such, these must be well understood and considered by the management.

6.7.1   Rules for Project Valuation

A project normally consists of the initial investment phase, followed by a series of years of operations, to end with a final year after which the project can still have residual value. For a good capital budgeting analysis, a set of rules that must be taken into consideration when building the cash flows to use for the decision-making process.

First of all, the project valuation must be based exclusively based on the quantitative analysis of *incremental cash flows*, namely, the additional cash flows to a firm generated by a project with respect to the situation ex-ante without the project. When deciding between two mutually excluding projects, only the differential cash flows of the two

Free cash-flow model	Operating cash-flow model
EBITDA	**EBITDA**
−Taxes	−Taxes
± Net working capital	± Net working capital
Net self-financing	*Net self-financing*
−Debt repayments	+ Operating divestments
+ Financial revenues	−Operating investments
−Financial charges	*Free cash flow to the firm*
+ Nondiscretionary divestments	
−Nondiscretionary investments	+ Nonoperating divestments
Free cash flow to the equity	−Nonoperating divestments
	−Debt repayments
+ Discretionary divestments	+ Financial revenues
−Discretionary investments	−Financial charges
+ New funding	+ New funding
−Dividends	−Dividends
Cash balance of the period	*Cash balance of the period*

FIGURE 6.6 Free cash-flow models.

projects are used. During the years of the project lifetime, other capital expenses can occur, for instance to restore the performance of the asset. The cash flows entail all the revenues streams yielded by the asset, the fixed and variable costs for the operation of the asset, maintenance, service, repairs and insurance payments, salvage values at end of asset life.

Opportunity costs, like the value of assets that can be employed differently (i.e., sold), must be taken into account and factored in the cash-flow estimates. For an asset that can be sold, its market value must be considered as a project cost.

Project *externalities* that can have an impact on other projects, like the cannibalization on existing products that are already in production and the sales of which will be reduced by the successful outcome of the project, must be reflected as well in the valuation model. Externalities can be very difficult to be singled out, but at least they should be considered in the project analysis to make sure that negative outcomes, if any, are either manageable or avoidable depending on the final go or no-go decision for the project. In this case, a thorough risk assessment as the one that will be outlined in the following of this chapter (see Sec. 6.11) is recommended.

Impacts from taxation, changes in the net working capital (equal to the changes in current assets minus changes in current liabilities), as well as changes in depreciation of assets can have a high impact on profitability. Let us consider, for example, an asset that replaces an older asset (that is sold) and the yearly depreciation of the new asset is higher than the old ones. The incremental cash flow that arises from such investment has simultaneous impact on depreciation and taxation. If the new asset has an annual depreciation of 110 monetary units (let us imagine a pure number value regardless of actual currency), while the old asset has an annual depreciation of 10 monetary units; the difference, 100 monetary units, will be the new depreciation amount to be subtracted to the cash flow according to Sec. 6.3.2. Such depreciation amount of 100 monetary units will be lower than the new asset depreciation of 110, causing therefore a reduced tax shield effect with lower taxes to be deducted from the yearly income statement. Of course, if the old asset is sold, the additional cash flow resulting from the sale is to be input to the cash flow of the year during which the sale takes place. The sale proceed itself will impact taxable income.

Revenue and cost streams that do not last for an entire year but cease to exist halfway through the same year (which is sometimes called *stab* year), must be curtailed by the amount that is not imputable to the full 12-month period. For example, state feed-in-tariffs that can be granted to the project as a subsidy and last until a date occurring before the last day of the year, must not be fully counted.

Since projects often deploy over a multiple-year period, changes and variations in macro-economical parameters must also be taken into account and, if possible, forecast with the help of specialized consultancies or internal research. Parameters that must be correctly considered in the valuation are, for instance, changes in inflation rates, interbank rates, foreign exchange rates, spread rates, hedging rates. Examples of other parameters that can change during the course of the project lifetime are the performance characteristics of the assets involved in the project. For instance, the reduction of the initial charge of an electrochemical battery depending on cycle numbers, the reduction in conversion efficiency of the power blocks. These reduction in performance are also linked to the necessary expenses for ordinary yearly maintenance to maintain performance itself, or the major refurbishments that are need to extend asset life over the project total life-cycle, that must enter into the cash-flow model.

Finally, if the assets that make up the project still retain value at end of period, or if the project itself is still operational and can still provide value over the next years, overall *salvage value* must be accounted for to cater for a correct valuation. In case end of period expenses must be incurred due, for instance, to decommissioning of the plants, disposal of self-generated waste or old equipment, shipment to waste landfills or dumps, these costs must be imputed in the cash flow of the last year to participate in the overall project value metric estimation.

Sunk costs are expenses that have already been paid and are not recoverable. The decision to proceed with the project or to stop it must be independent on such costs and must not be included in the capital budgeting cash flows. For instance, expenses for a consultancy that advise on the market potential of the outcome of the project, or the purchasing of the strings of macroeconomical parameters to be considered over the lifetime of the project, are costs that will not be recovered if the project is deployed; these costs for the consultancy will have to be added to the capital budgeting cash flows only if they are still to be made and will be paid as a part of the project development costs. Avoidance of sunk costs in project evaluation is a very important rule: sunk costs must not be considered to judge about the profitability of a project. Anyway, it is also important to understand that not necessarily all unrecoverable costs are sunk costs, as long as they can reduce the costs of the project once it receives the go decision; the purchase of an asset, not necessarily intended for such project, is not a sunk cost as long as such asset can be reemployed in the project itself, therefore, reducing the project overall outlays.

A summary of the basic rules to perform a correct project valuation are given in Table 6.4.

Consider:	
	Incremental cash flows
	Opportunity costs
	Impact on other projects (externalities)
	Impact on taxation
	Changes in working capital
	Changes in depreciation
	Changes in macroeconomical conditions
	Changes in project performance
	Major refurbishments of the assets
	Stab years
	Salvage value of the assets
	End of period mandatory expenses
Do not consider:	
	Sunk costs

TABLE 6.4 Rules for Project Valuation

6.7.2   Metrics for Project Valuation

When the incremental cash flows have been determined, a set of metrics need to be used to evaluate the profitability of a project, or rank projects when there are multiple investment opportunities.

Net Present Value

The *net present value* (NPV) is defined as:

$$\text{NPV}(r, n) = \sum_{t=0}^{n} \frac{X_t}{(1+r)^t} \tag{6.2}$$

where r is the discount rate, t is the period of time ranging from 1 to n, the number of periods along which the project unfolds, and X_t are the net cash flows for every t.

The discount rate is an important parameter that must be chosen carefully. It can be intended as the minimum rate of return, the hurdle rate, that could be earned on an investment characterized by a similar risk, or the opportunity cost of capital of the firm. The hurdle rate depends on the investor profile and can range between the return rate of low risk government bonds 3–4%, up to the returns required by venture capital investors, usually in the range of tens of percentage points.

For industrial or infrastructural investors, the hurdle rate is normally calculated as the *weighted average cost of capital* (WACC), namely:

$$\text{WACC} = k_e \frac{E}{D+E} + k_d \frac{D}{D+E}(1-T) \tag{6.3}$$

where k_e is the cost of equity, k_d the cost of debt, E the amount of shareholders' equity, D the amount of debt of the firm, and T is the taxation rate. Depending on the type of industry and macroeconomic conditions, WACC value can range from 4–6% to 12–14% or even more.

The NPV is a metric that embeds the concept of time value of money, meaning that cash flows that are far away in time are discounted more (value less) than cash flows that occur closer to the beginning of the investment period. The decision rule for the NPV is that, if the NPV is positive, the project can be financed; in case of multiple projects, the project with the highest NPV should be chosen.

One drawback is that the NPV has no benchmark to be compared against. The NPV is indeed a monetary value, therefore it should be used to compare only similar projects with similar initial capital expenses. If, for instance, Project 1 had an NPV of 1000 monetary units with an initial capital expenses of 1,000,000 monetary units, while Project 2 had an NPV of 100 monetary units with capital expenses of 10,000 monetary units, the management would need to assess which project to undertake according to metrics other than the NPV only. One key decision should be made on how to eventually invest the difference in the initial capital expenses, if such investments have some other benefits or strategic value (i.e., marketing positioning).

Internal Rate of Return

Another metric that is often used is the *internal rate of return* (IRR), which is the discount rate that equates Eq. (6.2) to zero:

$$\sum_{t=0}^{n} \frac{X_t}{(1 + \text{IRR})^t} = 0 \tag{6.4}$$

A project, to be allocated capital, should earn a higher than WACC return, since such projects provides a return higher than the minimum requested by the shareholders.

Apart from the difficulty in calculating the IRR iteratively, the IRR metric not necessarily exists as a single solution to Eq. (6.2) when the cash flows have a high degree of variability over the course of the project life; in such case, the NPV curve as a function of r can cross the zero line multiple times, making the use of IRR difficult if not impossible. Another difficulty in the use of IRR is the assumption that all cash flows are reinvested at the same IRR rate of the project, occurrence which is fairly difficult to meet especially when the IRR is high (and projects with similarly high IRR can be very difficult to find). IRR is then used extensively by private equity investors, since their capital allocation strategies entail a number of cash outflows and one or very few inflows at the end of the project life.

Modified Internal Rate of Return

To compensate for the assumptions on the IRR, a *modified IRR* (MIRR) metric has been devised where all cash inflows are capitalized to a future value at the last year of the project with a realistic reinvestment rate, the cash outflows are discounted to present value with a rate that is the rate applied to the investor from its financing partners (i.e., banks), and the MIRR is computed as:

$$\text{MIRR}(r, d, t) = \sqrt[n]{\frac{\text{FV}(r, n)}{|\,\text{PV}(d, 0)\,|}} - 1 \tag{6.5}$$

where r is the reinvestment rate, d is the financing rate, $\text{FV}(r, n)$ the capitalized cash inflows at r at time n, and $|\,\text{PV}(d, 0)\,|$ is the discounted cash outflows at time 0.

Pay-Back Period

The *Pay-Back Period* (PBP) is another very common and intuitive metric used extensively in industry despite of its inherent flaws. The PBP represents the number of years after which the sum of all yearly inflows outset the sum of all the yearly outflows. A project with a small PBP is usually considered as a good project since the money invested in the project will be paid back in a small number of years, and therefore receive the attention of some not-so-sophisticated decision-makers.

The PBP metric though does not embed the concept of time value of money, since inflows and outflows that happen several years in the future are treated the same as inflows and outflows occurring during the initial years. To consider the time value of money in the PBP calculation, a *discounted pay-back period* (DPBP) can be computed by using the discounted cash flows instead of the simple yearly cash flows as in the original PBP calculation.

Normally, when judging on a single project, if the analyst modifies some investment parameters (i.e., outflows like acquisition prices, costs of equipment, raw materials. . .) and gets a lower PBP, this can be a sign of the analysis moving toward a more profitable project. But when comparing two or more projects, a project with a lower PBP is not necessarily the best one. Imagine that Project 1 had a PBP of two years and a NPV of 10 monetary units, while Project 2 had a PBP of three years and a NPV of 1,000,000 monetary units. Which project would you choose? Savvy managers would likely go for Project 2. So, PBP has a second flaw, which is to prefer projects that has a lower period payback rather than projects that can be more profitable in a longer run.

Several examples of evaluation of projects can be found in the relevant literature. In Sec. 6.10 an example of a project valuation is provided as a quick reference.

6.8   Energy Project Metrics

When a project is developed, built, and enters operations, it can be embedded as an asset in a firm that manages that single asset or a number of other assets. Firms that exist as stand-alone organizations and that are embedding energy converting assets are known in the industry as *special purpose vehicles* (SPVs), where the purpose of the SPV is the development, construction and management, normally, of infrastructural assets. The structure of a SPV can be very simple (i.e., there are not HR, marketing, or sales departments), staff is kept to a minimum and usually all major operations are dealt with by external contractors.

Such form of incorporation is typical in the framework of project finance contracts (see Chap. 8, where SPV are also used to insulate the owners by the risks (especially financial) that are related to the project itself. This kind of organization is managed like all other firms; therefore a different set of metrics are needed to monitor if the assets are operated efficiently over their lifetime.

6.8.1   Return on Equity

One of the most frequently used metrics is the *Return on Equity* (ROE). It represents the ability of the firm to remunerate the equity and is given by:

$$\text{ROE} = \frac{\text{Net income}}{\text{Equity}} \tag{6.6}$$

The higher the value the more profitable has been the remuneration to the shareholders of the firm when all operations including taxation, have been carried through. For a profitable firm, the ROE must be higher than the hurdle rate of the shareholders of the firm.

The ROE depends on the industry structure: pharmaceutical industries will have different ROE than manufacturing firms, so one firm's ROE is usually compared against similar firms' ROE belonging to the same industry. An ROE lower than a risk-free interest rate is usually not a good sign of a profitable firm, since such firm does not manage to cover its operational risks with a sufficiently high return to its shareholders.

6.8.2 Return on Investment

Similar to ROE is the *return on investment* (ROI), defined by:

$$ROI = \frac{EBIT}{Total\ sources} \tag{6.7}$$

which expresses the profitability of the operations of the firm over the invested capital.

This is a simple and useful metric that can be used in combination with the ROE. In case the ROI is higher than the cost of debt, ROE would benefit from the investment and recurring to debt financing would improve the return for the shareholders, while the opposite occurs in case the ROI is lower than the cost of debt.

Both ROE and ROI are a ratio between a number in the I/S which represents a flow, and one in the B/S which represent a final status of the firm at the last day of the year. To improve precision, the average between the initial and last value of the equity and total sources is used in the denominators of Eqs. (6.6) and (6.7).

6.8.3 Economic Value Added

One metric that has recently gained in importance is the *economic value added* (EVA), which represents the actual value creation of a firm. This metric has been introduced in the early 1990s by the advisory company Stern & Stewart to compensate for the partial information provided by ROE or ROI metrics.

A positive EVA means that the firm is creating value for its shareholders. A simplified version of EVA can be defined as:

$$EVA = (ROI - WACC) \times Total\ sources \tag{6.8}$$

which shows that anytime the ROI is higher than WACC, the firm is creating value for the shareholders.

6.8.4 Levelized Cost of Energy

The *levelized cost of energy* (LCOE) [7] is the cost per kWh of energy computed according to the following formula (6.9):

$$LCOE = \frac{I_0 + \sum_{t=1}^{n} \frac{A_t}{(1+i)^t}}{\sum_{t=1}^{n} \frac{M_{el,t}}{(1+i)^t}} \tag{6.9}$$

where I_0 is the investment capital expense (or cost) at year 0, t is the year occurring during the asset lifetime of n years, A_t are the annual total costs, i is the interest rate (usually the WACC), and $M(el, t)$ is the yearly energy produced by the asset at year t. A_t cash outflows are computed according to Eq. (6.2), where costs can be normally identified as in Table 6.5.

The LCOE is an abstraction that represents the ratio between the discounted investment, commissioning, operations, and decommissioning cash flows of the asset as they variably materialize over time, over the discounted production of energy from the same asset.

Capital expenses (CapEx)
Energy storage
Power conversion
Balance of system
Control logic (software)
Operational expenses (OpEx)
Operations
Maintenance
Refurbishment
Depreciation

TABLE 6.5   Example of a Cost List for LCOE Analysis

The main advantage of LCOE is the fact that it grants the possibility to compare different energy technologies, provided that the assumptions are consistent in the comparison. LCOE can also be computed in real or in nominal terms, it is therefore mandatory to know if inflation has been taken into account in the cash flows used for its evaluation.

6.8.5   Profit Margin

The *profit margin* is a measure of how much net profit remains after all costs incurred by the firm to are taken out:

$$\text{Profit margin} = \frac{\text{Net profit } + \text{ Interest (net of taxes)}}{\text{Revenues}} \tag{6.10}$$

Typically profit margin is in the range of 5–10%, but it can also be higher than that depending on the type of industry (luxury goods and pharmaceutical companies often enjoying much larger profit margins), and on how good the firm is going. Many kinds of countermeasures can be taken by management to increase profit margins, like reducing costs or increasing sale prices.

6.8.6   Return on Assets

One metric that is very important in competitive environments is the *return on assets* (ROA):

$$\text{ROA} = \frac{\text{Net profit } + \text{ Interest (net of taxes)}}{\text{Average assets}} \tag{6.11}$$

The ROA provides a measure of how the firm is efficiently exploiting its assets. Assets that are not productively employed are just blocking capital that could be used in a better way.

An absolute number for ROA is not actually meaningful; it is much better to compare ROAs between companies in the same industry. Some specialized sources of information provide ROAs for many companies in different industries, but managers can analyze the financial statements of competitors to compute such values and get an idea of how good, or bad, their own assets are employed.

6.8.7 Current Ratio

The *current ratio* is a very useful, and used, metric to determine a firm's credit risk:

$$\text{Current ratio} = \frac{\text{Current assets}}{\text{Current liabilities}} \tag{6.12}$$

If the ratio is below 1, the firm is seriously facing the risk of not being able to cover its obligations in the short term. Normally, such ratio is above 2. A further refinement of the current ratio is the *quick ratio*, or *acid test ratio*:

$$\text{Quick ratio} = \frac{\text{Current assets} - \text{Inventory}}{\text{Current liabilities}} \tag{6.13}$$

which includes the inventory in the nominator to cater for the capacity of current assets to cover current liabilities in case inventory were to be considered illiquid in the short term.

6.8.8 Debt to Equity

The *debt to equity* (D/E) ratio is a very important metric used when management needs to understand the level of indebtedness of a firm:

$$\text{Debt to equity} = \frac{\text{Total liabilities}}{\text{Equity}} \tag{6.14}$$

If the amount of debt is very high with respect to the amount of equity, the firm can face problems with lenders in case revenues are not sufficient to repay all the debt obligations with respect to the lenders. For instance, if the D/E ratio is 3:1, 75% of the money used by the firm is provided by external lenders, or creditors. The higher such ratio, the less willing will the lenders be to provide additional debt due to the fact that too much risk is shifted to the external creditors.

In project finance (see Chap. 8), the debt / (debt + equity) ratio is used. In such cases, a high level of the ratio, normally around 80% is sought by investors to maximize their IRR. In project finance, a high level of indebtedness is normal since the projects can be structured to reduce the risk by off-taking the output of the project (i.e., sale of electricity) to public or big organizations with very low risk of bankruptcy.

6.8.9 Interest Coverage Ratio

The *interest coverage ratio* is another metric that can be used in combination with the D/E to analyze the liquidity situation of the firm:

$$\text{Interest coverage} = \frac{\text{EBIT}}{\text{Interest}} \tag{6.15}$$

If such ratio is lower than 1, it means that the firm can be facing issues in repaying the debt to lenders.

6.9   Nondeterministic Analysis

Each assumption that is embedded in the model is subject to some degree of uncertainty, which translates in the possibility for the parameters in the model to vary by assuming values in certain intervals [8]. Not considering the uncertainty in the assumptions makes the analysis limited and partial. A more refined analysis must therefore aim at reducing the uncertainty that is inevitably manifest in the decision-making process. A way to cope with uncertainty is moving from a deterministic towards a nondeterministic, stochastic modeling that is suited to deal with ranges of variation of the key project variables.

6.9.1   Sensitivity Analysis

A first step to deal with uncertainty is the use of the *sensitivity analysis*, which is the evaluation of how the model output (also referred to as the dependent variable) varies as a function of the model inputs (the independent variables). Sensitivity analysis is performed by varying the values of the inputs within their interval range, and computing the resulting values assumed by the model output.

Some inputs can have a large impact on the output even if varied by a small amount. These inputs are actually very important to be singled out during the decision-making process, most importantly if these small variations can have a high probability of occurrence. Regardless of their range of variations, all inputs that have a significant impact on the output are called the *high-sensitivity inputs*. Inputs with low impact are therefore referred to as *low-sensitivity inputs*.

When the inputs have been identified by their impact on the output, an important step is to understand the degree of control that the organization has over them. Such inputs can be divided into two further categories:

- the *endogen, or project-specific* input: the one over which the organization might have some degree of control and can, to some degree, have influence on its status; and

- the *exogen, or market-specific* input: the one over which the organization has little to no control and cannot influence its status.

Endogen inputs can be, for instance, the sizing of the energy system, the type of technology, the location, and contract structure, all under the control of the decision maker. Endogen variables can be placed under in control for example by negotiating liquidated damages in case of product performance failures (see Chap. 7), or by executing smaller scale projects before deploying the desired full-scale project.

Examples of exogen inputs are, for instance, the electricity prices, the regulations legally binding in a country, completely out of control of the organization.

It is important to understand that parameters that are high-sensitivity do not necessarily increase the uncertainty on the output; if such variables have a very low probability to change from their nominal value, they can actually have no impact on the project at all. Likewise, a low-sensitivity input can have a non-negligible contribution on the uncertainty of the output value if its likelihood of occurrence is very high. Therefore, the impact on the model output is a function of both the sensitivity and uncertainty of the model inputs. This means that when identifying the variables for the sensitivity analysis, we also need to understand the probability of their variation inside their interval. This is not always easy, and this entails the adoption of a stochastic approach. Sensitivity is therefore only the first step of the analysis.

6.9.2 Uncertainty Analysis

Uncertainty analysis is the evaluation of the impact of the variation of the inputs on the model output. Every input does not necessarily take on only one single number, but can accept a range of different values normally varying between a lower and a higher limit. The limit that reduces the likelihood to proceed with the project under analysis and makes the project output less appealing, is normally referred to as the worst case value, while the best case value is the one that achieves a better result in the output of the project and enhances the chances that the project can be judged feasible by the decision makers. A way to find the worst and best case values can be either resorting to experience or to take advantage of a statistical analysis on the input.

All parameters of a model assume a set of values in a range of variation with a certain distribution between certain limits. A way to describe how a parameter varies is by defining its *probability density function* (PDF), which is the function that gives the probability of a that continuous variable to assume a certain value, and is given by:

$$\Pr[a \leq X \leq b] = \int_a^b f_X(x)\,\mathrm{d}x \qquad (6.16)$$

where $\Pr[]$ is the probability, X the variable ranging within the limits a and b, and f_X the PDF.

For instance, if the variable is known to follow a certain PDF, the two or more boundary values can be defined as the ones providing the 95% confidence level that all other values are within the boundaries.

If the distributions of uncertainty ranges are defined for all the variables into consideration, a total uncertainty can be computed as a function of the uncertainties of each one of the variables. The estimation of the *total uncertainty* is provided by the formula by Kline and McClintock [9]:

$$U_R = \left[\left(\frac{\partial R}{\partial x_1}U_1\right) + \left(\frac{\partial R}{\partial x_2}U_2\right) + \cdots + \left(\frac{\partial R}{\partial x_n}U_n\right)\right]^{\frac{1}{2}} \qquad (6.17)$$

where U_R is the total uncertainty for a dependent variable R, x_i for $i = 1, n$ the independent variables, the partial derivatives are the sensitivities for each of the independent variables, and the U_i for $i = 1, n$ are the uncertainties of each x_i.

6.9.3 Monte Carlo Analysis

While sensitivity and uncertainty analysis certainly provide a method to analyze the impact of selected parameters on the final output, when such parameters become numerous, the evaluation performed one parameter at a time can be unfeasible or, at best, too time consuming.

The *Monte Carlo Analysis* (MCA) provides a way to overcome this problem by taking advantage of the computational capabilities of computers: a very high number of calculations are run, each one using different values for inputs taken within their probability distributions, and calculating their impact on the output. The output therefore assumes itself a probability distribution that is dependent on the probability distributions of the inputs.

An important caveat about the use of MCA is that the modeler must have a good understanding of the probability distributions of each parameter, the impact of which he wants to analyze: the quality of the modeling of the inputs will have a direct impact on the usefulness and validity of the results of the analysis.

All statistical PDFs can be used in a MCA; for instance, when a parameter is known to have a normal distribution with a mean and a standard deviation σ, a normal (or Gaussian) distribution is used and the MCA algorithm will randomly choose values within such distribution and calculate the corresponding output value.

The *normal distribution* describes the probability of occurrence a variable around its mean which is very common in many natural and non-natural processes and phenomena. It is a continuous symmetrical distribution of values always different from zero, with the PDF given by:

$$f_X(x) = \frac{1}{\sigma\sqrt{2\pi}} e^{-(x-\mu)^2/2\sigma^2} \tag{6.18}$$

where μ is the mean of the distribution of X values and σ is the standard deviation, given by:

$$\mu = \frac{1}{n} \sum_{i=1}^{n} x_i \tag{6.19}$$

and

$$\sigma = \frac{1}{n} \sum_{i=1}^{n} (x_i - \mu)^2 \tag{6.20}$$

When the variable is from processes that are skewed or always positive (or always negative), other distributions are better suited to describe the behavior of the variable. For instance, the *log-normal distribution* is the continuous probability distribution whose logarithm has a normal distribution. It is used to describe variables that are multiplicative products of independent random positive variables. The PDF for the log-normal distribution is given by:

$$f_X(x) = \frac{1}{x\sigma\sqrt{2\pi}} e^{-(\log x - \mu)^2/2\sigma^2}, \quad x > 0 \tag{6.21}$$

where μ is the log mean of the distribution and σ is the log standard deviation, given by:

$$\mu = \exp\left(\mu + \frac{\sigma^2}{2}\right) \tag{6.22}$$

and

$$\sigma = \sqrt{\exp\left(2\mu + \sigma^2\right) \exp\left((\sigma^2) - 1\right)} \tag{6.23}$$

The *Weibull distribution* is a continuous PDF which is extensively used in reliability engineering to describe failure mode patterns. The PDF for the Weibull distribution is

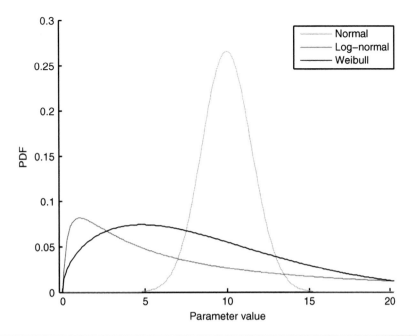

FIGURE 6.7 Example of normal, log-normal, and Weibull distributions.

given by:

$$f_X(x) = \frac{\beta}{\alpha}\left(\frac{x}{\alpha}\right)^{\beta-1} e^{-(x/\alpha)^{\beta}}, \quad x > 0 \tag{6.24}$$

where α is the *scale* parameter and β the *shape* parameter. The possibility to obtain different forms of distributions by simply changing the scale and shape parameters of the Weibull makes it a great PDF to model many possible behaviors of a variable.

The normal, log-normal and Weibull's distribution PDFs are shown, for comparison, in Fig. 6.7, where the normal distribution has a mean of 10 and a standard deviation of 1.5, the log-normal has a mean of log 10 and standard deviation of 1.5, and the Weibull's has scale 10 and shape 1.5.

The *Beta distribution* is a continuous probability distribution that is defined on the interval [0,1] by:

$$f_X(x) = \frac{x^{\alpha-1}(1-x)^{\beta-1}}{\int_0^1 u^{\alpha-1}(1-u)^{\beta-1}\mathrm{d}u} \tag{6.25}$$

where α and β are the shape parameters, and the function at the denominator is the *Beta function* that normalizes the PDF within the [0,1] interval. The Beta distribution is used in statistics to model the behavior of random variables that are limited between a minimum and maximum value. This makes the Beta distribution very useful to model variables in the MCA that do not extend outside the interval of definition. Figure 6.8 shows a family of Beta distributions according to different shape parameter values.

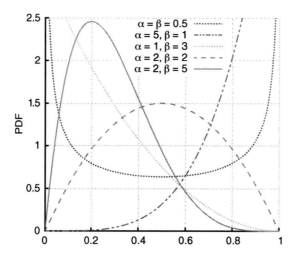

FIGURE 6.8　Example of the beta distribution at different values of the shape parameters [10].

Another useful distributions with many potential applications in MCA is the *uniform distribution*, where a parameter has the same probability to take on each single value in its full range of variation.

Not all distributions are necessarily continuous. Many parameters that have impact on the output of a project model are discrete by nature and cannot be modeled with continuous PDFs. For instance, events that can happen on a distribution line, like interruptions on the supply of electricity to an industrial customer. A *discrete distribution* can therefore be constructed by deciding the values assumed by the discrete random variable, and the respective values of the probability of their occurrence, making sure that total probability will exactly sum up at 1.

When analyzing discrete distributions, the concept of probability density function ceases to exist because the underlying distribution is not continuous. In this case, the *Probability Mass Function* (PMF) is used instead, where a PMF is the function that yields the probability that a discrete random variable takes on a precise value.

For a discrete random variable $X \subseteq A$, where A is a subset of $\mathbb{D}$, the set of discrete numbers, the PMF is the probability that $X = x$, $x \subseteq A$ given by:

$$f_X(x) = \Pr(X = x) \tag{6.26}$$

with:

$$\sum_{x \subseteq A} f_X(x) = 1 \tag{6.27}$$

The PMF graph for a discrete distribution is shown in Fig. 6.9. The values taken on by the discrete random variable are 1, 4, 7, and 10 having each a probability of occurrence equal to 0.2, 0.5, 0.1, and 0.2 that, summed up, equal to 1.

Another example occur when considering three (or more) different scenarios, in which a variable can take on specific values in each one of the scenarios. For instance, electricity price forecasts are usually provided as strings over a period of time (normally 10 or more

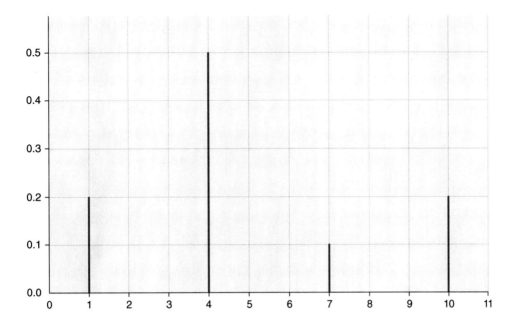

FIGURE 6.9 Example of a PMF graph of a discrete random variable with values 1, 4, 7, and 10, each with a probability of 0.2, 0.5, 0.1, and 0.2.

years) in a high, central, and low scenario, each one built over different assumptions on the future of the electrical grid and markets (see Sec. 5.8). In this case, a *triangular distribution* can be used, distribution that is constructed by evaluating the probability of occurrence of the values that the parameters hold in the worst case, likeliest case, and best case scenario.

When events can occur independently from one another and the average number of such events is known, the *Poisson's distribution* can be employed. The PMF for a Poisson's distribution with average value λ is given by:

$$f_X(x; \lambda) = \Pr(X = x) = \frac{\lambda^k e^{-\lambda}}{k!} \tag{6.28}$$

The PMF graph for a Poisson's distribution is shown in Fig. 6.10.

Many other PDFs and PMFs can be employed in an MCA analysis, depending on the degree of precision and applicability as decided by the modeler. Several plug-ins or stand-alone software are available to perform a good MCA on the impact of inputs in the model of the project under scrutiny.

6.9.4   Real Options

A *real option* is the value of the investment that gives the possibility to perform a later, bigger, even riskier investment in a real asset or real project [11–13]. Many projects can be evaluated using real option techniques. The rationale is that a project can have value even if it is not actually deployed in its full entirety. The value of the option lies in the fact

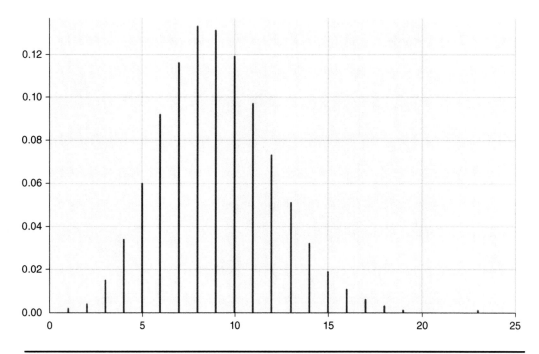

FIGURE 6.10 Example of a PMF graph of a discrete random variable with a $\lambda = 9$ Poisson's distribution.

that the decision maker can later have the possibility to continue or abandon the project, if the outcome of the initial investment (i.e., the option) proves to be, or not to be, of sufficient worthiness. In some instances, the evaluation of a real option is an important step that provides the decision maker with the added value of managerial flexibility. In other words, the real option is the value of the possibility, but not the obligation, to perform a subsequent investment postponed in time.

The valuation of the option is performed through the use of the *Black–Scholes option pricing formula*; initially devised for financial options evaluation, the formula has been extended to the valuation of real assets. The formula is the following:

$$V_{opt} = S e^{-dt} N(D_1) \; - \; X e^{-rt} N(D_2) \tag{6.29}$$

where

$$D_1 = \left[\ln(S/X) + \left(r - d + \frac{1}{2}\sigma^2 \right) t \right] \Big/ (\sigma \sqrt{t}) \tag{6.30}$$

and

$$D_2 = D_1 - \sigma \sqrt{t} \tag{6.31}$$

with

- V_{opt}: the value of the real option
- S: the present value of the underlying project cash flows

Black–Scholes option pricing model

Inputs:

Stock price, S	700,000.0	Present value of future cash flows.
Strike price, X	1,25,000.0	Present value of future initial costs.
Volatility, σ	50.0%	
Risk-free rate, r	3.0%	
Time to expiration, t	2	
Dividend yield, d	4.0%	The annual costs to maintain the option.

Outputs:

$D1$	−0.49472	
$D2$	−1.20182	
$N(D1)$	0.31040	
$N(D2)$	0.11472	
Call price	**65,530.5**	The value of the option.

FIGURE 6.11 Example of evaluation of a real option using the Black–Scholes formula.

- X: the present value of the initial cost of investing in the project at the future date
- r: the risk-free interest rate
- d: the rate of the annual cost of maintaining the option (if any) over S
- t: the time during which the option is valid, namely how long the decision to invest will be delayed
- σ: the volatility, in %, of the project cash flows
- $N(\bullet)$: the cumulative distribution function of a normally distributed variable

Figure 6.11 provides an example of the calculation of a real option using the Black–Scholes's formula. If the investment to enter the option costs less than the value of the option, the decision maker can reasonably entering the option and be ready to initiate the future investments when more information is available.

This method of computing the value of the real option is based on a set of financial assumptions that do not translate perfectly when it comes to the evaluation of real projects. The analogous of a real option in finance, is the *call option*, which is the right and not the obligation to purchase a financial asset in the future. One of the underlying assumptions in a real option is that it does not take into consideration the fact that the project could not be feasible t years from now due to, for example, changes in laws, regulations, or market conditions. In a trading analogy, it would mean that we will not have the possibility to sell the underlying stock due to the absence of a market for it. Nonetheless, the Black–Scholes formula can be used as a first approximation since its closed form makes it easy and quick to apply. Other techniques have been developed to provide better estimates of real options, often depending on the type of industry to which they relate (i.e., pharmaceutical).

6.10   Example of a Project Valuation with Monte Carlo Analysis

An example of a project valuation using the techniques described in the previous sections is provided in this paragraph; the model has been simplified a bit, the numbers changed, but the framework shown is from an actual real-life energy storage project for an industrial application.

The main sources of revenues come from three storage applications. The most important revenue contribution comes from the avoidance of the interruptions that occur on the grid distribution line that is supplying electricity to the industrial load. The industrial customer is accepting to be interrupted without notice, but is rewarded a fixed amount of money per MW of interruptible power by the utility. Depending on the actual number of interruptions, the customer has to pay back, or receive, a payment depending on whether the number of interrupts is under or above 10 per year. Further revenues are obtained by the use of the ESS for peak-shaving and time-of-use applications.

The simplified P&L and cash-flow statements are given, for the first years, in Fig. 6.12.

The number of interrupts are modeled as an input to the MCA according to a Poisson's distribution. Since historically the industrial customer is experiencing an average number of annual interrupts as high as nine, the Poisson's λ is equaled to 9 and the PMF is the one shown in Fig. 6.10. All interruptions are considered as occurring independently from one another.

The revenues from peak-shaving are modeled as a beta distribution with the two scale parameters at 1 and 0.75, and the minimum and maximum set at 7.5 and 9.2 (see Fig. 6.13).

The revenues from time-of-use of energy are modeled after a Weibull distribution with scale at 2 and shape at 10 (see Fig. 6.14).

The result from the MCA for the IRR computed over a 10-year period are shown in Fig. 6.15; the IRR has an average of 7.9%, with a 90% confidence level of having an actual IRR between 6.3% and 9.5%. In a conservative analysis, the decision maker should consider using 6.3% in his budget.

A sensitivity analysis on the three input variables is provided in the tornado diagram in Fig. 6.16. The most impactful variable is the number of interrupts requested by the utility. This is clearly an exogenous parameter on which the decision makers have no power to influence whatsoever.

6.11   Risk Assessment and Response

An often overlooked but extremely useful and much needed tool to minimize the chances to fail, *Risk Assessment and Response* (RAR) is a process that must absolutely be performed for every project. Too many projects have been lost because of risks that were not considered at the beginning of the development and, when manifest, they were not coped with. Risk comes from the absence of full knowledge on the many factors correlated with the outcome of the project, and causes an undesirable outcome to all or part of the project. RAR attempts to mitigate the negative outcome of all risks involved in all phases of the project.

RAR is a four-step process that can be referred to as IQAR, from the initials of the four phases: *Identification, Quantification, Analysis, and Response.* The first three steps collect,

Year	0	1	2	3	4	5	...
Inflation factor	1.00	1.01	1.02	1.04	1.05	1.06	
Energy inflation factor	1.00	1.01	1.02	1.03	1.04	1.05	
Electricity yearly multiples	1.00	0.96	0.94	0.87	0.85	0.88	
ESS performance decay		1.00	0.99	0.98	0.97	0.96	
Year from start		1	2	3	4	5	
Year of operation		1	2	3	4	5	
P&L ('000)	**0**	**1**	**2**	**3**	**4**	**5**	**...**
1 – Capacity payment (NOT indexed)		147.0	147.0	147.0	147.0	147.0	
2 – Peak shaving (indexed)		8.6	8.6	8.6	8.6	8.6	
3 – Time of use (indexed)		18.3	17.9	16.6	16.1	16.8	
Other revenues		0.0	0.0	0.0	0.0	0.0	
Total revenues		**173.9**	**173.4**	**172.1**	**171.7**	**172.3**	
O&M		9.9	10.0	10.1	10.2	10.3	
Initial ESS charge		1.0					
Insurance		3.0	3.1	3.1	3.1	3.2	
Management fee		0.0	0.0	0.0	0.0	0.0	
G&A		1.5	1.5	1.6	1.6	1.6	
Other costs		0.0	0.0	0.0	0.0	0.0	
Total costs		**15.4**	**14.6**	**14.8**	**14.9**	**15.1**	
EBITDA		158.5	158.8	157.4	156.7	157.2	
D&A		103.7	103.7	103.7	103.7	103.7	
Interests earned/paid on DSRA		0.0	0.0	0.0	0.0	0.0	
Interests		61.0	53.1	44.8	36.2	30.4	
EBT		-6.2	2.1	8.9	16.9	23.2	:
IRAP		2.4	2.4	2.4	2.3	2.4	
IRES		2.0	2.1	1.8	1.7	4.8	
Other taxation							
Net income		**-10.6**	**-2.4**	**4.7**	**12.9**	**16.0**	**...**
Cash flow ('000)							
+ Net income		-10.6	-2.4	4.7	12.9	16.0	
+ D&A		103.7	103.7	103.7	103.7	103.7	
+ interest (sponsor loan)		51.3	46.7	41.6	36.2	30.4	
+ Add. debt		0.0	0.0	0.0	0.0	0.0	
- DSRA		0.0	0.0	0.0	0.0	0.0	
- Rep. senior debt		0.0	0.0	0.0	0.0	0.0	
- Rep. add. debt		0.0	0.0	0.0	0.0	0.0	
- Add. assets		0.0	0.0	0.0	0.0	0.0	
Free cash flow to the Equity	**-1,036.5**	**144.4**	**147.9**	**150.0**	**152.7**	**150.0**	**...**

FIGURE 6.12 Extract of the first years simplified P&L and cash-flow statements for the project.

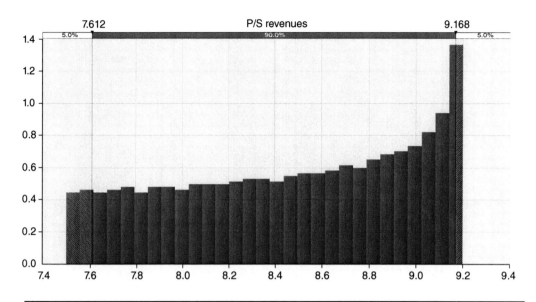

FIGURE 6.13   Input of the MCA for the peak-shaving revenues in the project.

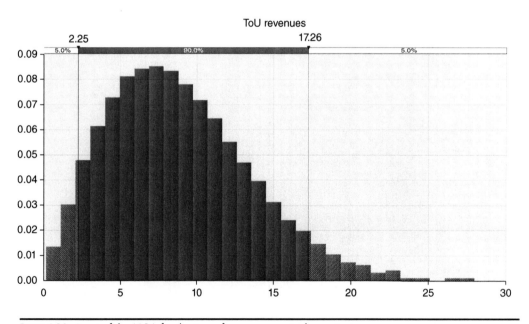

FIGURE 6.14   Input of the MCA for the time-of-use revenues in the project.

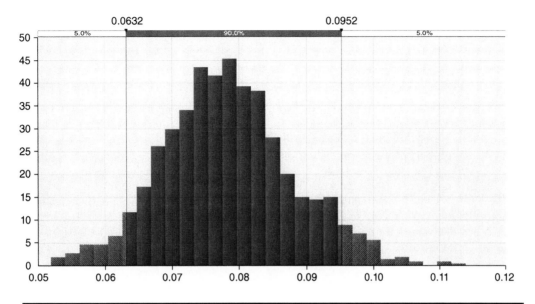

FIGURE 6.15   Output of the MCA for the IRR of the project.

analyze, and quantify the risks, while the final fourth is the response, the actual and factual way that the project owners use to deal with the occurrence of risk both preventively and if should happen.

6.11.1   Identification

Identification is the determination of all significant risks and their causes that are likely to occur throughout the life of the project.

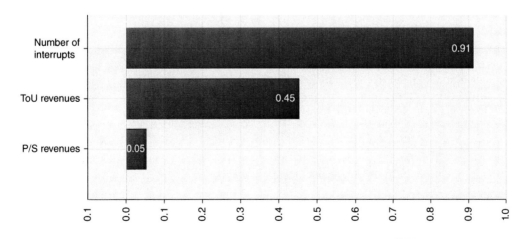

FIGURE 6.16   Sensitivity on three inputs, tornado diagram.

The techniques to perform the identification phase are based on knowledge of the project environment and are best performed through informed brainstorming sessions. Before taking on the identification brainstorming session, the team should be knowledgeable about what they will talk about (the more knowledge, the lower the uncertainty). Nonetheless, the team should comprise of people with different point of views and experiences.

The risks identified in this stage should not have overlapping boundaries, but taken together should be close to represent the whole spectrum of events that could occur, they should therefore constitute a set of *exclusive but exhaustive* items.

A difficult part of the identification phase is considering the combination of events that, while taken singly would not cause issues to the project, if taken together could synergize and have major impact on the project. The team analyzing the RAR should be, therefore, very wary about all interconnections of potential problems and brainstorm to understand and list them.

A possible way to try to cover all possible risks is to make use of tables where broad areas of risk are already outlined. Possible areas for risk identification are:

- cost
- environment
- facilities and equipment
- materials
- market
- organization
- people
- resources
- scope
- technology
- time

6.11.2   Quantification

Quantification is the gathering of all possible information on the potential outcomes for the risks singled out in the identification phase. This entails reaching a quantitative knowledge about their probability and distribution to get an understanding on their potential effects on the project.

Quantification is based on the following two main parameters:

- the damage (D_{risk}) that the risk is capable of causing to the project should it materialize, and
- the probability of occurrence (P_{risk}) of the risk

Both must be quantified and the values consolidated together in order to compare them. The mathematics used to evaluate P_{risk} and D_{risk} can be defined in many ways, but must be maintained consistent all along the analysis. A possible way to quantify P_{risk} is by using tables similar to the following Table 6.6:

Probability	Value
0–25%	1
26–50%	2
51–75%	3
76–100%	4

TABLE 6.6 Possible Definition of P_{risk}

If the risk-assessment team considers a probability of occurrence of 30% for a specific risk, then the corresponding value 2 will be chosen as the quantification for P_{risk}. Probability can be obtained either by historical or statistical analysis if data are available, or by approximation if data are not available or too time consuming, too expensive, or impossible to retrieve. In the construction of tables like the one in the previous example, it is important that at least four ranges are used in order not to have analysts crowding all their evaluations in the middle of the scale.

If damages cannot be quantified analytically, D_{risk} can be defined similarly to P_{risk} by using tables, while some forms of weighing can be included to enhance disruptive risk from being masked by low P_{risk}. One way is to categorize damage in a 1–5 scale by qualitative reasoning and use an exponential function with that value as exponent (see Table 6.7):

To consolidate the numbers, P_{risk} and D_{risk} can be multiplied together to obtain the value that will be associated with that specific risk; these values constitute the quantified relative weight of the risk.

6.11.3 Analysis

Analysis is the use of the knowledge derived to categorize the potential problems according to their likelihood and impact on the project. The most important tool of all is without doubts the *Pareto analysis*, based on the principle that 80% of the effects come from 20% of the causes.[2] By classifying the risks according to their respective relative weights, it is possible to single out the risks that have high potential impact on the project and prepare a response strategy to those risks depending on their relative importance. The 20% risks having the highest impact can be classified as *Class A* risks, to be taken care of with the utmost care and priority, while the remaining risks can be classified as *Class B* and *Class C* risks according to their decreasing impact on the project.

An example of an IQA for a real energy storage project is given in Fig. 6.17 (not all items are shown).

---

[2] Vilfredo Pareto (1848–1923) was an Italian engineer and economist that made important contributions on the allocation of resources and distribution of income. He introduced, for instance, the concept of *Pareto's optimality* (or *Pareto's efficiency*), which is the state in which it is not possible to change the allocation of scarce resources to a number of individuals without making one better off at the expenses of another individual that would be worse off.

Damage	Value	Exponential
Low	1	e^1
Mild	2	e^2
Medium	3	e^3
Severe	4	e^4
Disruptive	5	e^5

TABLE 6.7 Possible Definition of D_{risk}

6.11.4 Response

Response is the determination of the best course of action to address each risk according to its likelihood and impact. The response strategies must be identified in advance in order to perform a quick, agreed-upon, efficient, and timely action before or when the risk hits. A set of possible response categories is the following:

- contingency planning
- risk assumption
- risk avoidance
- risk mitigation
- risk prevention
- risk transfer

Contingency planning is the definition of the course of action to be taken when the risk materializes. Contingency plans must be devised in advance, optimized and, if possible, practiced during the course of the project. For instance, simulated drills by involving all necessary resources and individuals can even be considered in case the risk is affecting not only the economics of the project but also the safety of individuals.

Assumption of risk is the acknowledgment that a risk is identified, but no response or action is taken nor any contingency planning is devised. The risk is accepted as it is and the project will bear the consequences should such risk occur. The rationale behind such decision can be due to a very low impact of the risk, or the costs to take action against it are simply higher than its benefits.

Avoidance is the action of eliminating the exposure to the risk. For example, if some conditions occur, the project stops or is delayed.

Mitigation of risk's impact is the response that aims at reducing the effects on the project. An example of a mitigation strategy is a countermeasure that protects when the risk materializes, like adding sprinklers to reduce the damages to premises in case of fire. Many risk responses belong to this kind and can be compulsory by law when the impact is not only on material assets but also on people.

Prevention puts the project in the condition to reduce or eliminate risk's probability of occurrence. This should be the first step to take to face high-impact risks. An extreme risk-prevention strategy would even consist in the decision not to pursue the project and abandon it altogether.

Identification		Quantification			Analysis
Area	Event	Probability	Damage	P*D	Pareto
Market	Variable prices	4	5	594	
Market	Customer exiting or closing down	4	5	594	
Legal	Law forbidding some sources of revenues	4	5	594	Class A
Legal	Change in law	4	5	594	
Market/legal	Tax uncertainty and missing regulations	4	5	594	
...		...	...	...	
Cost	Technology costs too high	3	5	445	
Technology	Technology not suitable for different customers	3	5	445	
Organization	Joint ventures with market complementors	3	5	445	Class B
Organization	Only one technology provider	3	5	445	
Market	Only one dominant source of revenues	3	5	445	
...		...	...	...	
Market	Hard to patent the technology	4	4	218	
Market	Low entry barriers for competitors	4	4	218	
Market	Hard to forecast energy prices	3	4	164	
Environment	Force majeure	1	5	148	
Market	Financing? Bankability?	2	4	109	
Market	Market saturation	4	3	80	Class C
Organization	Insufficient internal resources	4	2	30	
Market	Vale proposition difficult standardise to many different customers	4	2	30	
Cost	Initial development costs unknown	2	2	15	
Technology	Difficult to system integrate	1	2	7	
...		...	...	...	

FIGURE 6.17   Example of output for IQA steps.

Transfer is a common way to address a risk before it materializes: insuring against it by having someone else bearing it, at a cost. The organization is, therefore, accepting to pay an amount today to hedge against the eventuality to pay higher costs should the risk occur. The insurance company will bear all or part of the consequences of an unfavorable outcome.

6.11.5   Continuous Risk Management

Even in case of a very thorough and comprehensive RAR exercise, many risks are simply impossible to identify or forecast. This means that the organization must develop a mindset that is capable of producing quick and efficient responses to the threats that can come in the way, and maintain a proactive dynamic control of the risks that can develop during project's lifetime.

Regular updates of the RAR process must be performed and eyes kept open during all phases of the project.

References

1. Friedlob G. T., Schleifer L. I. F. (2003). Essentials of Financial Analysis. John Wiley & Sons.
2. Brealey R., Myers S., Allen F. (2013). Principles of Corporate Finance. McGraw-Hill.
3. Damodaran A. (2010). Applied Corporate Finance. John Wiley & Sons.
4. Brealey R., Myers S. (2003). Capital Investment and Valuation. McGraw-Hill.
5. Koeller T., Goedhart M., Wessels D. (2010). Valuation: Measuring and Managing the Value of Companies. McKinsey & Co.
6. Vance D. E. (2003). Financial Analysis and Decision Making. McGraw-Hill.
7. Kost C., Mayer J. N., Thomsen J., Hartmann N., Senkpiel C., Philipps S., Nold S., et al. (2013). Levelized cost of electricity. Renewable Energy Technologies Study. Fraunhofer ISE.
8. Electric Power Research Institute (2004). Energy Storage Technology Valuation Primer: Techniques for Financial Modeling. EPRI, R. Schainker (Ed.). Palo Alto, California.
9. Kline S. J., McClintock F. A. (1953). Describing uncertainties in single-sample experiments. Mechanical Engineering 75:3–8.
10. Wikimedia, author Horas based on the work of Krishnavedala, date 14 november 2014, web link commons.wikimedia.org/wiki/File : Beta_distribution_pdf.svg.
11. Black F., Scholes M. (1973). The pricing of options and corporate liabilities. Journal of Political Economy 81:637–654.
12. Morton R. (1973). Theory of rational option pricing. Bell Journal of Economics and Management Science 4(1):141–183.
13. Villiger R., Bogdan B. (2006). Pitfalls of valuation. Journal of Commercial Biotechnology (12)3:175–181.

CHAPTER 7

Business Law Concepts

Summary. This chapter provides the reader with an essential knowledge of the main concepts of business law applied to energy projects. The description of the agreements and contracts is based on actual documents used with projects that have been actually built or transactions that have been fully executed.

7.1 Introduction

A certain degree of knowledge of the basis of business law is fundamental to achieve proficiency in the management of energy projects [1]. Alongside with good negotiating capabilities, a sufficiently good grounding in how to read contracts and the capability to interact with lawyers during the different stages of the project implementation can increase the chances to effectively lead a sound business proposition to a successful closing and transforming an idea into reality.

It is of course out of the scope of this chapter to cover all the intricacies of business law and the many difficulties that can be successfully coped with only by experienced lawyers, whose help is always needed especially in large and costly transactions that involve multiple business partners. Still, not having a clue of what is going on during the long sessions between lawyers can be a problem, since many times the lawyers themselves need to resort to management decision or professional help to side-step nontrivial impediments during the negotiation. Being in line with what they do and understanding what is going on is essential to make things happen.

This is why this section is meant to provide the reader with a view on the main concepts of business law applied to energy projects. The descriptions of the agreements and contracts all come from real-life deals and actual documents used in projects that have been actually built or transactions that have been fully executed, whether they were development, construction, operations, or acquisition projects.

7.2 Legal Systems

Different legal systems exists around the world and it is very difficult to generalize and have a comprehensive view of all legal systems enforced in all countries, but a as a reference, there are basically three legal systems that can encompass the majority of them:

- civil law
- common law
- religious law

whereby some countries enforce a pluralistic legal system by combining civil law and common law in their jurisdictions, or even mixing civil with religious laws.

Civil law is the most widespread of all legal systems. It is based on written codifications collected in books, called *codes* which are issued and, in case, amended only by governmental bodies that have legislative power to issue and enact laws; codes are interpreted by the jurisdiction and judges have no power to enforce new rules or change the codes.

A first historical example of law codes dates back to the Code of Hammurabi (18th century B.C.), but the first widespread adoption of such law systems is due to the Roman empire, which embedded previous civilizations and religious laws with non- or partially coded behavioral social decision-making. Expansion of the Roman empire, the establishment of the Franks' empire in central Europe with their code-based *Lex Salica* during the early Middle-Ages and later influences of France under Napoleon's empire, spread the adoption of such legal system in many countries in continental Europe, South America, Africa, and Asia.

Common law is a legal system that was developed in England and Wales in the Middle-Ages, influenced by anglo-saxon cultures and spread by the expansion of the British Empire. In this system, alongside with a set of laws and statutes issued by legislative bodies, decisions are taken by judges and written in the form of *precedents* which are subsequently used by other judges for decision making in similar cases. Precedents, therefore, accrue over time to form a large body of legal knowledge, where each precedent is structured in the form of the *holding*, namely the principle that can be drawn by the opinion of the court, and the *obiter dictum* which is a nonbinding remark or opinion incidentally made by the judge.

Within Common law is the *Equity* legal system, which was developed by the Court of Chancery in England to reach decisions in equitable interests, where equity is a set of principles constituting what is fair and right as per a *natural law* concept. In Equity, the judge decides based on principles of fairness and justness. Starting from the late 19th century, a series of *judicator acts* issued by the English Parliament aimed at fusing the two different systems without reaching a complete merge between the two.

Religious laws are either intended as moral guidance or directly implemented in the judicial system of a nation. Examples of religious laws are the *Canon law* for Christianity, *Sharia* for Islam, and *Halakha* in Judaism.

7.3 Agreements and Contract Formation

A contract that is entered into between two or more parties is a legal entity that, to be valid and enforceable in front of the society, needs to be respectful of a set of determined prerequisites and embed those in its formation. A *contract* is a promise, or a set of promises, made between two or more parties to perform specific courses of actions that are duly enforceable in front of the law, and the breach of which can be remedied by the law. The parties are usually referred to as the *offeror* and the *offeree*, the latter being the entity to whom an offer to enter into a contract is made by the offeror. For a contract to be valid, binding and enforceable, it must usually embed three conditions:

- the mutual assent between the parties
- the mutual consideration between the parties
- no defenses against the enforcement of the contract

The *mutual assent* is manifest in the contract if: the parties express a *commitment* to perform something specific; such performance must meet with criteria of *certainty and definiteness*, with clearly defined conditions of quantity, quality, terms, prices, and delivery valid under the local commercial codes; a written *communication* of the offer to the offeree; the presence of *termination* clauses that can be by act of the parties, through options or by law; a written *unequivocal acceptance* by the offeree.

The *mutual consideration* is expressed by a *bargained-for-exchange* where the offeror promises to do something in consideration of the offeree promising to do something in exchange, and the *balance between benefit and detriment* whereby the offeree is giving up something of defined value (the detriment) to the offeror who receives it (the benefit).

The contract must bear *no defenses* against its enforceability, examples of this being the following: it must be in writing[1]; do not have *unconscionability* against the protection of the weaker part from excessive unfairness or contract asymmetry; must not be have defenses based on lack of capacity of one party.

If all conditions for a contract formation are fulfilled, the parties are met with the *sanctity of contract*, where the parties cannot escape from its obligations unless under the conditions explicitly conceded by the law.

There can be cases in which the contract is nonperformable and its nonperformance do not give rise to liabilities on the parties. Impossibility or impracticability to perform a contract are referred to as *frustration*, which is a legal concept that gives way to terminate contracts when unforeseen events make contractual obligations impossible to be performed or radically change the ground purposes that motivated the parties to enter into the contract. Within frustration: *hardship* occurs when for unforeseeable circumstances the contract balance becomes too burdensome for one party and hardship clauses are designed to provide the parties with the possibility to renegotiate in good faith the contract; *force majeure* is, for instance, environmental, political and similar events leading to potential contract disruption as described in more detail in the following Secs. 7.9 and 7.10. Such nonperformability of contract can be, for instance, due to a *change in law* that occurs after the contract has been signed and that causes some assumptions on which the contract was construed to be illegal or causing the same contract to be illegal.

If a contract is *null*, it means that it is invalid from the beginning, while legal actions have to be taken in order for a contract to be judged *void*.

7.4 Breach of Contracts and Remedies

A *breach of contract* occurs when there is an absolute duty to perform but the duty is not discharged by the breaching party. A *minor breach* occurs if the affected party still gains the substantial benefit of the transaction since the minor breach can be, for instance, a small delay in completing performance or small deficiencies in the quality and quantity of the goods. The remedies are therefore intended only to provide a compensation to the affected party while not taking out the duty to perform from the breaching party. A *material breach* instead is a failure to perform by one party that causes the aggrieved party to suffer serious damages or losses and the affected party does not receive the

---

[1] The *Statute of Frauds* is a Common Law instrument which requires that certain kinds of contracts must be memorialized in signed writing.

substantial benefit from the failure to perform of the breaching party. The affected party have the right to terminate the contract and/or ask for immediate remedies. A worst-case scenario of a material breach is the *fundamental breach*, or *repudiatory breach* so that the affected party can proceed in suing the breaching party for the nonperformance. Finally, an *anticipatory breach* is the declaration by one of the parties stating the intention to not fulfil its obligations under the contract.

Following contract breach, the affected party can ask for a *legal remedy* (or *judicial relief*), whereby a jurisdiction enforces a right of the affected party to protect from the damages or losses it suffers as a consequence of the breach.

Legal remedies provide the breaching party with the faculty to seek from the non-breaching party a set of damages, usually correlated with the disbursements of specific amount of money from the breaching party to relieve for losses or undue gains. Such damages can be categorized into:

- compensatory damages
- liquidated damages
- nominal damages
- punitive damages
- restitution damages

Compensatory damages are classified into "expectation," "reliance," and ``consequential'' damages. *Expectation damages* are sufficient for the affected party to purchase or acquire a substitute performance; *reliance damages* are intended to grant the nonbreaching party the recovery of the costs for the nonreceived performance on the part of the breaching party as an alternative to expectation damages if too difficult to be determined; *consequential damages* are designed to protect and cover any further losses that result from the breach and that are foreseeable at the time of contract signing.

The *liquidated damages* (LD) are based on provisions that have been agreed to by both parties at the time of contract. They are intended to compensate and not to punish the breaching party (in which case it will be referred to as a penal or penalty clause), are set at a level that is based on reasonable forecasts of the damages that the nonbreaching party will suffer, and are to be paid when the event occurs whether or not the nonbreaching party actually suffers money or pecuniary losses.

Nominal damages are awarded when there are no actual losses but the breach is shown.

Punitive damages, also called *exemplary damages* are generally not enforceable in commercial contracts except under limited or very peculiar circumstances.

Restitution damages are awarded to the nonbreaching party and entail the breaching party to give up the gains received from the execution of the contract by reimbursement of such gains to the other party.

In Common law jurisdictions, legal remedies include also a set of *equitable remedies*. The two most common equitable remedies are: *injunctions*, which can be "mandatory" by requiring a party to do something, or "prohibitory" by having a party not to do something; *specific performance*, which requires a party to perform a course of action like honoring a contract.

Another type of remedy is the *declaratory relief*, where a jurisdiction determines the rights of the parties to an action without awarding damages or ordering equitable remedies.

Normally, in many contracts a clause is stated as the *inadequacy of damages* as the following:

> Without prejudice to any other rights or remedies that the parties may have, the other party acknowledges and agrees that damages alone may not be an adequate remedy for any breach of the terms of this agreement by the other party. Accordingly, each of the parties may be entitled to the remedies of injunction, specific performance, or other equitable relief for any threatened or actual breach of the terms of this agreement.

granting the access to equitable remedies alongside that to damages.

The use of some of the remedial measures described in this section will be shown in the following chapters.

7.5 Dealing with Risks

Business ventures often entail an enormous amount of risk, as well as create an enormous amount of opportunities. To make sure that the benefits from the exploitation of such opportunity be maximized and the risks associated with it be minimized, a long and articulated way of coping with business risk has been developed over time to protect against unfavorable outcome during the different phases of the negotiation, construction, and operations of the assets.

7.5.1 Clauses and Provisions

Ways to ensure that one party is actually getting what expected are secured by resorting to the following clauses and provisions that constitute a set of potential countermeasures to confront with business risk:

- *Call-put options*: are courses of action where one party has the right, but not the obligation, to perform. A call option is for instance, when one party can acquire an asset (i.e., by buying), a put option when a party can alienate an asset (i.e., by selling).

- *Cherry-picking* of assets: if the local jurisdiction so permits, the acquiring party can choose to proceed with closing only on a portion or subset of the assets, in case they can be divided in parts and if this grants leaving liabilities to the left-over assets

- *Closing accounts* analysis: is the evaluation of all assets and liabilities of the firm based on its financial statements prepared in accordance with the accounting principles by the party owning the firm, at a defined date and drafted on the basis of the definitive accounting data duly reported on the accounting books of the firm. Also called *final accounts*.

- *Conditions concurrent*: are events that must occur together and that the parties are bound to perform concurrently.

- *Conditions precedent*: are events that must happen or done before some act-dependent thereon is performed or a deal closed.

- *Conditions subsequent*: are events that brings an end to one party's legal rights or duties.

- *Earn-out*: is a variable part of price that can be paid at the end of the interim period at closing or after certain tests or performances mutually agreed-upon or after a determined period of time if certain conditions are met.

- *Guarantees*: issued by parent companies, corporate companies, banks, insurance companies or financial institutions, usually with a requested acceptable rating (i.e., S&P's ratings in the A range).

- *Indemnity protections*: are a form of compensation that entitles the acquiring party to recover from the other party a particular loss that has been suffered. For instance, indemnity protections can be cash payments, replacements, or repairs to be performed by the selling party. One advantage of an indemnity protection is the fact that the acquiring party does not need to actually show that it has suffered a loss.

- *Price reduction*: when one party deems that the acquisition has a diminished value than what previously thought to the event leading to such diminishing; asking for a price reduction is one potential request to the selling party that can still ensure that the transaction is not terminated but consummated at closing date.

- *Price retentions*: for a period of time by the use of, for instance, *escrow accounts*, which are temporary bank or other financing entity accounts where the funds of a transaction are placed out of the control of the two parties, but where the money is freed and transferred if some conditions precedent are verified by a third party (i.e., a notary). This allows one party not to part with funds representing its assessment of the amount at risk.

- *Remedies*: see Sec. 7.4.

- *Staggered completion*: if the closing can be performed separately after the outstanding obstacles standing in the way are cleared, payment of the price can be staggered accordingly.

7.5.2 Representations and Warranties

An important protection for both parties is the *representations and warranties* (R&W) section, which is always present in business contracts and can take several pages with declarations by one of the party that, if not true, have legal consequences on the mendacious party.

Representations are declarations by the parties asserting facts that are true at the time the representations are made. Warranties are declarations that guarantee the other party about some future state occurrence and give rise to potential requests for indemnity in case the assertions are false. Often, there is no clear distinction between the two. R&W are meant: to provide a cause of action to enable recovery by a party if a representation or warranty is untrue and one party is damaged as a consequence of such invalid representation or warranty; to force disclosure by the other party during the due diligence process so to receive more accurate information before signing, therefore taking certain actions (like, for instance, price decreases) as a consequence of information that reduce the value of the project for the business of the acquiring party.

Usually, the selling party will want to provide few and specific R&Ws, to be given at signing and with no adjusting mechanisms, especially on price. The more specific the

R&Ws are, the less likely the risk to incur in liabilities on selling party's side. The acquiring party on the contrary will seek to obtain a long and comprehensive list covering all possible eventualities in the transaction. One way for the acquiring party to enforce the R&Ws is to have the selling party to repeat the same R&Ws at closing after having already stated them at signing.

Some important representations and warranties are the *due organization* status, where a party represents itself as being

> . . . a company duly organized, validly existing and in good standing under the law of incorporation and has all requisite power and authority to own and operate its business and properties and to carry on its business as such business is now being conducted and is duly qualified to do business in their respective place of business and in any other jurisdiction in which its performance of the agreement makes such qualification necessary. . .

the *due authorization* of a party which:

> has full power and authority to execute and deliver the agreement and to perform its obligations hereunder, and the execution, delivery and performance of the agreement by it have been duly authorized by all necessary action on its part. . .

and under the *no insolvency* representation, upon which the party

> . . . is not insolvent or in bankruptcy or in similar circumstances, is neither bankrupt nor subject to any insolvency proceedings or similar circumstances under the Laws applicable to it according to the country of its incorporation.

The *noncontravention* states that the incorporation and by-laws of the party are not in contrast with the execution and performance of the services in the agreement, and the party has a commercial interest in pursuing the consummation of the agreement.

The party also warrants the *continuing representations*, where all representations and warranties

> . . . will remain true, valid and correct throughout the term of duration of the agreement and undertakes to keep the other party fully indemnified and harmless from any damages, costs or expenses suffered by the latter as a result of a breach by the party (or its affiliates) of any of its obligations as above or, otherwise, from any of these representations and warranties to be incomplete, untrue, not correct or not valid to any extent.

7.5.3 Due Diligence

Before a deal is signed, the parties engage in a performing a thorough and comprehensive *due diligence* (DD) process. The goal is the disclosure of all liabilities of the target company. Such liabilities can be depending upon outstanding issues with local governments, taxation problems, employment or social security contributions, missing authorizations, wrong book-keeping, or similar issues. In case of projects in which investors or other financial entities decide to provide their equity or debt, some checks can be required from investors to advisors in order to gain insights or confidence before continuing the investment process. Legal, accounting, and technical consultants are hired to analyze the project status and compile signed due diligence reports. The selling party must give access to a full set of confidential documents that are collected in a *data room* and, in case, provide an access period as agreed in appropriate legal agreements that are enforced before signing of the final contract (see Sec. 7.7). Some liabilities that can arise during the DD process can become *deal breakers*, which are liabilities that irreversibly modify the original set of information used by the acquiring party as the assumptions on which the business case was made and numbers run (i.e., if shares are pledged, assets less valuable than expected, litigation with third parties).

A *legal due diligence* entails mostly the analysis and validation of the permitted process that has led the SPV or the firm to hold legal and valid titles to the realization of the energy plant and its actual construction and operation on the market. Analysis on permission documents and contract validity, surface rights, easement rights, past history of ownership titles, and notary checks on plots of lands, are between the activities performed during a legal DD.

A *technical due diligence* covers all technical aspects peculiar to the project: electrical design, mechanical design, performance evaluation, energy productivity, compliance between permitted design versus as-built, positioning of the plant over the correct land parcels as secured by land surface or easement rights contracts, analysis of EPC and O&M contracts and their comparison with industry standards, verification of price and terms of payment in comparison with market standards, checks during construction, and assistance during acceptance tests and other ancillary activities.

An *accounting due diligence* is performed by accounting firms to provide the investors with a detailed view of the financial situation of the SPV or the organization that owns the energy plant. Aspects covered by their DD can be the analysis of the balance sheets, profit and loss statements, accompanying notes, the evaluation of the net financial position (NFP), assets, and debt tax situation of the vehicle under scrutiny by the investor.

It might also happen that *ad personam* checks can be performed under request. When hiring a top executive or when engaging in negotiations with potential sellers or developers of a project, the need to have total certainty that the people involved in the transaction are not or have not been involved in fraudulent activities can be considered as necessary.

Particular care should be posed when relying on legal, technical, and accounting advice if the investors lack the internal capacity to double check the information provided by the advising teams. For this purpose, often the investors request a *reliance letter* to be issued by the advisors or a formal collateral warranty to protect over the consultant's appointment. The need to be able to rely on the report produces the consequential need to establish a cause of action in case the information provided by the advisors were misleading or incorrect and caused a damage or loss to the investing party.

Possible reactions to the information that is disclosed during a DD process can be: proceed with further investigation, usually extending the period of exclusivity that can be linked to the DD process; negotiate a price reduction; proceed with indemnities; request the selling party to fix the problem; and use of escrow accounts.

7.5.4 Interim Period

A typical deal procedure at this stage is where one party, the purchasing or acquiring party, have received from the other party a certain set of financial, technological, commercial, information; therefore, a price and outline terms for the transaction completion (or *closing*) have been agreed by the parties, the deal is signed and the position of the target asset and the parties is warranted by the selling party. The time period between signing and the date of closing is referred to as the *interim period*. The risk at this stage of the transaction is borne by the acquiring party either because it is not able to recover losses by bringing up a claim against the other party, or because it is not able to refuse to close the deal. If a transaction is signed and closed the same day, which is the case of many transactions, the acquiring company can seek to recover the loss, if suffered, on the basis of a breach of the R&W. Larger deals of complex transactions can entail conditions to be satisfied during the interim period and before closing occurs. Covering the risk can therefore be more difficult and special arrangements and provisions might need to be enforced. A typical course of action, in case of a material breach by the selling party, is for the acquiring party to be able to not perform the transaction and reach closing.

Provisions to cover risk during the interim period between change of ownership from one party to another can be dealt with by resorting to temporary agreements like the ones described hereby. An *interim period negative control* clause that seeks to ensure that the business is carried out as per the ordinary course and that no out of the ordinary commitments are made without acquiring party's consent or knowledge. Negative control is intended as the right of the acquiring party to refuse the performance of some undertakings by the selling party to avoid losses in the value of the company being managed in the interim period. One potential issue is that the acquiring party can incur in liabilities due to its managing the company if the interim period starts taking too long, handover is delayed, and decisions have to be made to maintain company's value. Another potential issue is related to limit company's value by wrong negative control decisions that might interfere with the correct course of business for the to-be-acquired company. The *material adverse change* (MAC), or *material adverse event* (MAE), clause is designed to enable the acquiring company to pull out in case of events that are capable to strongly affect the business during the interim period and reducing the asset's value. Parties affected by a MAC can request a price revision in case of acquisition of the asset, or get out of the transaction altogether. This clause is often entered upon request of financing parties in project finance or mergers and acquisitions transaction that usually take a long period of time before closing.

7.5.5 Insurance

To protect against risks during an asset construction or its day-by-day operations are the *insurance* contracts, entered into by an *insurer* party and an *insured* party and normally designed to hedge against specific risks. In exchange for an initial payment (the *insurance premium*), the insurer accepts to pay the insured party in case the latter suffers damages or losses.

A specific terminology has been developed in close combination with the mechanics of insurance policies.

Material losses are damages or losses that affect the cost of the asset (goods, plants, real estate, machinery...); they are a *direct material loss* if the loss is originated on the asset by the direct effect and action of one or more of the events insured by the policy, or a *consequential, or indirect, material loss* if the event has had a inevitable chain of consequences that have led to other losses after the action on the asset (i.e., the loss of production due to an event that has damaged a manufacturing line).

Immaterial losses are the damages that affect the value rather than the pure cost of the asset; they can represent a *direct immaterial loss* if the event on the asset has caused a damage not proportional to the cost of the asset (i.e., the loss of valuable data in a data center that is well above the value of the machinery constituting the data center itself), while *indirect immaterial losses* are all losses not related to the previous definitions (i.e., a damage to an industrial building can stop all activities within, creating the conditions to incur in additional losses like loss of profits from the impossibility to proceed with the operations.

Not necessarily all insurance policies cover all these for different types of losses, this is why several different insurance policies can be entered into by the insured parties. Therefore, insurance contracts are designed to address very specific issues and make them different from other types of contracts by introducing peculiar concepts and mechanisms. Usual types of insurance contracts are hereby described.

Liability insurance is a class of compulsory insurances for insured parties at risk of being sued by third parties for negligence or wrongdoing. There are three forms of liability insurances, public, product, and employer. *Public liability* insurances are intended to protect from impact on third parties like workers, contractors,. . . ; *product liability*, to hedge against problems due to manufactured products distributed and sold on markets; *employer liability*, to cover liabilities on an employer from its own employees. Examples of liability insurances are the following.

A *general own civil liability* insurance covers the losses for material or personal damages and their relevant consequences that occur against the parties working directly for the insured as a consequence or in relation to the activities or services of the insured. A *general third party civil liability* insurance covers the losses for material or personal damages and their relevant consequences that occur against third parties as a consequence or in relation to the activities or services of a party. Between the coverages are warranties for civil liability for accidental pollution and contamination. An insurance against *civil liability for the operation of vehicles and machinery* for all vehicles subject to compulsory insurance pursuant to local regulations. An insurance against *accidents at work* that covers accidents during the works for all employees of the contractor or of subcontractors.

Other kind of insurances are entered into to provide covering from damages and losses sustained by the business.

An *all risks property and business interruption* insurance covering direct physical loss or damage to the plant and associated property including risks associated with business interruption. Such insurance is normally written on a full replacement cost basis for new property. Direct and indirect losses are covered by this policy.

An *erection all risk* (EAR) insurance covers direct physical loss or damage to the assets and associated property including all materials, supplies and equipment installed or to be installed during the works. Such insurance is normally written on a full replacement cost basis for new property.

The *advance loss of profit* insurance covers the losses due to delays in the works and the postponement of the project being able to provide revenues to the owners.

A *marine and land transit* insurance, covering any and all materials and equipment (including spare parts) while they are in transit to the site of the projects with sum insured for not less than the full replacement of the transported goods value included freight costs from supplier's warehouse to client's premises or project site.

Most insurance policies introduce constraints to the mechanism of loss reimbursements. Two main provisions are the *deductible*, which is the part of the damage that remains on the insured party responsibility and is not reimbursed by the insurance company, and the *maximum indemnity coverage* (or *indemnity limit*) which is the maximum damage that can be covered by the insurance policy.

Some other provisions can be found in contracts like, for instance, the following hereby described.

The *leeway clause* provides the insured party that, in case of incorrect statements in the original policy agreement and the total final insured sum be higher than the one originally insured, insured amount will be automatically increased after payment of the additional premium if such increment does not exceed a percentage, usually, of the 20% of the total initial insured sum.

A *backstop guarantee* is a guarantee from a third party (the *backstop guarantor*, usually banks or insurance companies) ensuring the coverage of such obligation in case the original obligor does not or cannot comply with it.

The *waiver of subrogation* is a clause that prevents the insurer from obtaining restitution from a third party who causes any kind of loss to the insured. The client and eventual financing parties shall be released from any responsibilities and the right to ask the two parties for compensation be waived, both by the insured party and by the insurance company, in connection with the risks covered by the signed insurance policies.

The *loss payee* clause (also known as *loss payable* clause) is common in insurance contracts when the asset is financed by a third party; the clause provides that in the event of payment being made under the policy in relation to the insured risk, such payment will be made to the financing entity and not the insured beneficiary of the policy.

The *cross liability* clause provides that the policy will be intended and in force as if such policy had been issued to each of the insured in their own right, so that the interest of each coinsured will not be invalidated or detrimentally affected by any act or omission (including a breach of warranty or policy condition) of any coinsured. This clause ensures that if a coinsured is actually causing the damage, the damaged party under the same insurance will still receive coverage. Each party under the same insurance will be treated as a separate entity even if under the same policy.

7.6 Nondisclosure Agreement

A *nondisclosure agreement* (NDA) is entered into between two parties that need to avoid improper disclosure of sensitive information that can damage the competitive positioning of the parties. Usually, an NDA is the first step between parties that are willing to enter in negotiations for the construction or acquisition of projects.

The agreement begins with the *by and between* section that gives all the necessary information on the parties that are entering into the agreement, followed by a preamble, the *whereas* section that states why the parties are entering into such agreement, normally referred to as *the purpose*. As a consequence, the *disclosing party* is the entity that discloses confidential information to the *receiving party* that receives confidential information from the disclosing party. In cases where the NDA must protect a two-way information flow, both parties are defined as receiving or disclosing and the NDA is therefore *mutual*.

Confidential information is defined as:

> ...any information or document (containing economic, financial, technical, commercial, strategic or any other kind of information regarding the business, transactions, products, technology, or any other content of relevance for the Disclosing Party) that is by any means provided (orally, in writing or recorded in any kind of support) at any date, either before or after the execution of

this Agreement, by the Disclosing Party and/or its Representatives to the Receiving Party and/or its Representatives. Further, it will be deemed to be Confidential Information the observation of tangible objects such as prototypes, samples, products, and facilities, including, but not limited to, trade secrets, know-how and other intellectual or industrial property or information relating to the Disclosing Party's business, operations, products, technology, together with any and all analyses, compilation, study, summary, extract, drawings, or other documents of any kind prepared by either Party, or by both of them jointly, or any of their Representatives that contain or otherwise reflect any Confidential Information disclosed by either Party.

Such confidential information cannot include all information that has been already made available to the public domain, already known by the receiving party, is made available by a third party without breaching the NDA with that third company or is independently developed by the receiving party. The *limited Warranty* clause states that such confidential information is provided "as is."

The core of the agreement is the *nondisclosure* clause:

At all times, the Receiving Company shall treat and hold the Confidential Information in secret and as confidential information and shall not, and shall cause its Representatives not to, disclose, either orally or in written form, the Confidential Information of the Disclosing Company to third parties or to the Receiving Company's own Representatives, except to those Representatives who meet the requirements. For these purposes, the Receiving Company shall restrict to the minimum extent the number of people who have access to the Confidential Information and to comply with the provisions of the applicable laws relating to personal data protection.

If the receiving party is forced to disclose the confidential information due to a legal or judicial obligation (the *exceptional circumstances* or *compelled disclosure* clause) must be notified to the disclosing company so that it may take appropriate measures to prevent such disclosure or to waive the application of the NDA, but without impeding the receiving party to comply with its obligation versus the legal or judicial requests. The *maintenance of confidentiality* clause requires that the receiving party take at least those measures that it takes to protect its own most highly confidential information to protect the secrecy of and avoid disclosure and unauthorized use of the confidential information of the disclosing party.

Commercial clauses are also enforced in order to avoid the entering into a binding transaction agreement; the *no obligation* clause states that nothing obligates either party to proceed with the purpose or any other transaction or relationship between them, and each party reserves the right, in its sole discretion, to terminate any discussions contemplated in the NDA. The *exclusive ownership and no license* secures the ownership of the confidential information to the disclosing party, and nothing should be intended as a change of ownership of the information as a consequence of having shared that information with the receiving party.

Limited representations and warranties provides a minimum amount of representations of the parties like their duly incorporation and good standing in their respective jurisdictions, that they have full authority to enter into the agreement so to make it valid, legally binding, and enforceable.

The article about *remedy for disclosure* acknowledges that damages would not be an adequate remedy for any breach of the confidentiality and the disclosing party may restrain, by equitable relief like specific performance or injunction, any conduct in breach of the agreement. *Term* and *survival* clauses provide the time limits of the nondisclosure validity, with survival being the remaining time after expiration of the agreement (defined in the term clause), during which the rights and obligations of the parties remain in validity.

Other clauses about the *return of materials* specify how to deal with the mechanics of the restitution or destruction of the tangible or intangible support that is containing the confidential information and the applicable *governing law and jurisdiction*.

Confidentiality is also requested to avoid disclosure of the existence of such agreement to third party (*publicity*) and to avoid letting someone else know of the discussions that are taking place concerning the potential or existing business relationship in connection with the purpose of the NDA. Companies might anyway need to disclose if required by applicable law or any listing or trading regulations concerning its securities, provided that the receiving company gives the disclosing company prompt written notice of such requirement and, if needed, collaborate to seek for the right to protect the confidential information from public disclosure.

A set of clauses is needed to regulate some meanings or mechanics of the NDA. The *integration* clause states that the NDA is the complete agreement that supersedes prior oral or written agreements. The *construction* clause is an acknowledgement of the status of both parties who enter the agreement as conscious sophisticated partners of equal bargaining power and agree that this agreement shall be construed as if jointly prepared and drafted by both Parties and that under no circumstances will any provision hereof be construed for or against either party. The *counterparts* clause states how many copies of the NDA must be exchanged in original signed form. The *severability* of the agreement is a recurrent clause in many other types of contracts or agreements and indicates that if one or more of the provisions are found to be invalid by jurisdiction, such provision will not affect the validity of any of the remaining provisions. If *amendments* are needed on the NDA, the same modifications will be made in writing and signing. The *waiver*, another common clause in many agreements tells that a

> ...failure to exercise a right or remedy, or such Party's acceptance of a partial or delinquent payment, shall not operate as a waiver of any of such Party's rights or the other Party's obligations under this Agreement and shall not constitute a waiver of such Party's right to declare an immediate or a subsequent breach.

The *assignment and successors* and *no third-party beneficiaries* clauses provide the NDA to be transferred to other entities without prior written notice, but the agreement is passed onto the successors and assignees, while no other parties except the disclosing party, the receiving party, their successors, and assignees will have any direct or indirect cause of action or claim in connection with the NDA. Furthermore, the *relationship* clause clarifies that having entered into this agreement neither signifies that the parties have entered in any kind of business venture together nor that any party incurs in any obligation on behalf of the other party.

A usual final clause gives the mechanics to notify all communications related to the NDA (*Notices*).

An extension to a NDA is the clause of *noncircumvention*, where the recipient party agrees that it will not contact the ultimate owners of the assets or attempt to circumvent the disclosing party or its agents in any way. Any violation of this covenant shall be deemed an attempt to circumvent the providing party, and the recipient shall be liable for damages in favour of the providing party, which may also apply for injunction relief as the damage may be difficult to assess or may be irreparable.

A typical structure for an NDA is represented in Table 7.1:

Main body:	
	By and between
	Whereas
	Recitals and definitions
	Limited warranty
	Nondisclosure
	Exceptional circumstances
	Maintenance of confidentiality
	No obligation
	Exclusive ownership and no license
	Limited representations and warranties
	Remedy for disclosure
	Term
	Survival
	Return of materials
	Governing law and jurisdiction
	Publicity
	Integration
	Construction
	Counterparts
	Severability
	Waiver
	Assignment and successors
	No third-party beneficiaries
	Relationship
	Notices
Annexes	

TABLE 7.1 Typical Structure of an NDA

7.7 Letter of Intent, Memorandum of Understanding

A *letter of intent* (LOI), or *memorandum of understanding* (MOU), is a document that records a preliminary understanding between parties who intend to take some actions leading to an outcome of interest for both.

A first LOI between the parties is an *assurance LOI*, which has the purpose to state the seriousness of an intention to advance in further negotiations between the parties, and can be requested by third parties such as prospective lenders, financial institutions, and board of directors.

After advances in the contacts between the parties, a *framework LOI* can be signed to set the next steps in the milestones in negotiations and the identification of the agreements to govern the transaction process but without undertaking of obligations at the expirations of the deadlines. At this stage of the negotiation rounds, some initial binding commitments can be entered into by the parties like, for instance, for the protection of trademarks, the confidentiality agreements, the express denial of agency or partnership, eventual cost allocations and joint efforts (like when entering codeveloping projects or other forms of joint ventures). A legal consequence of this is to proceed with further negotiations, but with no commitment to enter into a final agreement.

In complex negotiations, a *memorialization LOI* can be signed subsequently to record preliminary or partial agreements, with the main purpose of setting a moral commitment on which the parties should not go back. Here, the legal consequence is that, since many material elements are still to be determined, the parties are still not bound by a final contract. Some grey areas still exist, like no proper R&W, price setting, good faith or best effort obligations, and standstill agreements.[2]

Three possible effects of an LOI can be the following:

- No legal effect: the wording used in the LOI are not "agree" but rather "understand" or "intend," and phrases like "subject to final contract."

- Partially enforceable: with expiration dates, confidentiality agreements, standstill agreements or exclusivity clauses and breaches protected by reference to equitable remedies and/or liquidated damages, and the *authority to proceed*, where one party may start performing before the execution of a final contract, and with some form of compensation damages in case of failure to enter into a final agreement.

- Contracts to negotiate: set a standard of conduct and undesired behavior, like the refusal to negotiate or improper tactics like, for instance, extreme inflexibility, unreasonable proposals or abrupt withdrawals from the negotiation tables.

Depending on the type of LOI, its construction can take a rather informal form with no article-like structure, or a more articulated, contract-style document. The beginning paragraph states the interest by one of the parties over a purpose (or *intent*) and puts forward that the LOI is intended to serve as the basis for further discussions and negotiations between the parties with respect to the purpose. To assert that the LOI is not intended to be a binding contract, the letter states so by adding that

> ...any further binding commitment will arise only upon the negotiation, execution, and delivery of mutually satisfactory definitive agreements and the satisfaction of the conditions set forth therein.

A request to proceed with a formal set of further actions can also introduced in the LOI. For instance, the start of a due diligence process can be requested as beginning from the execution date of the LOI, whereas during such period the parties are considered to be entering an exclusivity period, usually of the duration of the due diligence process itself. Based on the preliminary information, and subject to the eventual results of the due diligence, the LOI gives an indicative price or economic consideration that will regulate the proceeding of the works toward reaching the purpose of the LOI. An LOI can anyway maintain its nature of a *nonbinding offer*, but in case some provisions inside the LOI are

---

[2] A *standstill agreement* is signed when the parties agree not to deal with other third parties for a period of time.

binding, usually non-binding and binding provisions are separated in two different parts of the LOI.

A "Part One" can report provisions on the transaction, on the purchase price, and conditions for the acquiring party to proceed with the acquisition of the project. A miscellaneous or ancillary section may introduce the requirement for the parties to

> ...make comprehensive and thorough representations and warranties to the Buyer which would survive the Closing, and would provide comprehensive covenants and other protections for the benefit of the Buyer, including indemnifications of Buyer against losses arising out of misrepresentations and undisclosed preclosing liabilities.

``Part Two" can describe the binding provisions of the LOI, like an exclusivity clause granted by the selling party to the buying party and an eventual break-up fee in case of breach of exclusivity, provide access to a suitably arranged data room where all the relevant documents to analyze the transaction are kept and updated on a regular basis, confidentiality agreement, and disclosure of information pre- and postclosing.

An *access period* can be agreed upon between the parties whereby

> ...during the period from the signing date and the termination date, the Seller will afford the Buyer and its representatives, full access to the Target Company, its operations, personnel, customers, vendors, representatives, properties, contracts, books and records, and all other documents and data.

A *best efforts* clause can be envisioned for the parties to

> ...negotiate in good faith and use their best efforts to promptly arrive at a mutually acceptable Definitive Agreement for approval, execution, and delivery on the earliest reasonably practical date.

The letter can end with details on the sharing of costs, governing law, and terms of the agreement, like the end date after which the LOI is not to be considered valid any longer.

7.8   Project Development Agreement

A *project development agreement* (PDA), or *project collaboration agreement* is entered into by parties willing to codevelop a project by sharing costs, revenues, or equity ownership of the project. It is basically a letter of intent stating details about the project and how to proceed with its development and management of the special purpose vehicle that will own the project.

After the usual "by and between" and "whereas," the "purpose" sections, emphasis is placed on the respective tasks to be accomplished to reach the purpose. Only to provide an example for the "scope of works" section, due to the many different kind of projects that can be covered by a PDA, one party can take care of the site identification, analysis and selection, negotiation of surface or easement rights, the initial feasibility study and analysis of current legislation, the environmental impact analysis, the geological and geotechnical analysis of the site, the design and permitting for the connection to the electric transmission or distribution grid, and the construction of the financial model to assess the profitability of the investment. The other party is usually the contractor that will proceed with the construction of the project; this party normally takes care of the design for permitting (preliminary design) and construction (final design) for all civil works, electrics, and industrial automation, in general all hardware and software of the project. Each party will internally bear the costs of these activities, while legal, advisory, notary expenses are normally shared between the two.

If an SPV is incorporated to own the project assets, the PDA apportions the SPV shares to each party according to negotiation, while if permitting of the project will be granted by the public authorities, the construction and maintenance of the assets is reserved as a *right of first refusal*[3] to the construction party.

Normally, the agreement is completed with the usual clauses about confidentiality and industrial property, term, assignment, and arbitration. In case a *limitation on liability* clause states that no party shall be obligated for any debt, encumbrance, liability of the other party solely by reason of the agreement.

7.9   Engineering, Procurement, and Construction Contract

An *engineering, procurement and construction* contract (EPC) regulates the relationship between the client and the contractor for the design, the purchase of all components, structures, electrics, civil works, in general all the technology needed, the construction and commissioning, including the performance tests, of the project.

The contract starts by stating the date in which the agreement is signed and the "by and between" section that gives all the necessary information on the parties that are entering into the agreement, followed by a preamble, the "whereas" section that summarizes the boundary conditions that are leading to the signing of the agreement. For instance, what the two parties are doing as a business concern, if there are preliminary contracts signed by the parties and directly related to the EPC contract, a brief description of the energy project with the main technical details (i.e., power, capacity, location. . .), if the project has been duly authorized and a statement similar to the following:

> . . . the Parties, mutually acknowledging sufficient legal capacity to contract and oblige, as well as to the representatives of the Parties sufficient legal authority and faculties to represent them, freely agree to execute this Agreement, upon the terms and conditions set forth hereto. . .

representing the mutual consent outlined in Sec. 7.3.

The *recitals and definitions* paragraph precisely defines, in alphabetical order, the main terminology recurring in the agreement, for instance terms such as "authorization," "business day," "civil code," "date of execution," "performance," "plant," "price," and "service." A set of articles in the paragraph are reserved to describe the mechanics of the use of adverbs ("hereof," "herein," and "hereunder" that are intended to refer to the whole agreement and not a portion of it, or names, words, the heading of the articles, references to laws, statutes, regulations, provisions, all concepts that must be taken into account to avoid misjudgment and misinterpretation when reading the contract.

A paragraph lays out the *terms of the agreement* to set the conditions precedent that must be respected in order for the contract to become valid in full force and effect.

---

[3] The *right of first refusal* (RFR or ROFR) is a contractual right similar to a call option, giving its holder the option to enter a business transaction with another party before such party is entitled to enter into that transaction with another third party.

The *scope of works* paragraph states that:

The purpose of this Agreement consists of the design, supply, construction, assembly, commissioning, and delivery by the Contractor to the Client of the Project, on a turnkey basis. The Client hereby entrusts to the Contractor, which accepts, the execution of the Project in consideration for the Price. The Project will be completely finished, tested and ready for commercial operation upon the agreed terms, pursuant to the provisions of the applicable Law and this Agreement (and particularly in accordance with the terms of the Technical Specifications and the Final Design.

The services provided by a contractor to a client in the framework of an EPC agreement totally depend on the type of the project, but can be generalized as the following:

- *Engineering.* The design and supply of preliminary, final and as-built schematics, site layout, design calculus, technical reports, operation and procedure manuals for maintenance, component and product certificates, their test reports and quality control checks.
- *Civil Works.* The supply of all materials, equipment, machinery, and labor needed for the construction of the foundations, buildings, accesses, drainage, fencing, voltage transformation and grid-connection cabins.
- *Electro-mechanical equipment.* The supply and installation of electrical cabling, overcurrent protections, overvoltage protections, earthing systems, electrical cabins or substations, protection interfaces for LV, MV, and HV lines, field cabins comprehensive of indoor ancillary services (ventilation, cooling, UPS systems).
- *Monitoring, control, and automation System.* The supply and installation of metering units for the measurement of main plant parameters (power, energy, and weather stations), data-loggers, malfunction alarms, communications bus, industrial automation systems (PLCs, industrial PCs).
- *Surveillance system.* The supply and installation of anti-intrusion and antitheft systems like dome cameras, motion-detection systems, infrared cameras, fiber-optic cablings.
- *Construction, setting-up, assembly, and installation.* The management of all loading, transportation, unloading, warehousing and security of all the equipment, tools, utilities, machinery, and material supplied.
- *Prevention of labor risks.* The contractor must comply with all current regulations concerning labor and environment health and safety on the work site.
- *Works management.* The supervision and management of the works of the project in accordance with the industry standards and take the obligations and responsibilities established under current legislation.
- *Management and carrying out of tests.* The management and execution of the electrical, mechanical, technical, operational, and guarantee tests to ensure the full, correct, and safe functioning of the project.
- *Connection to the grid.* All the required activities and works for the connection of the project to the grid.
- *Permitting.* All the necessary permits that must be complied with for the correct and safe execution of the project.

The *changes and exclusions to scope of works* articles ensure that: no changes shall be effective without the written agreement of the parties, usually in the form of an amendment to the EPC contract; if changes become necessary, they should be considered as extra works and the contractor will charge the extra works to the client; the client is not to unreasonably withhold or delay acceptance of the changes are provided with similar level of warranties and conditions to the initial works in the scope of works of the agreement. Exclusion clauses are entered to avoid that the contractor could build or supply any works, equipment, and services not expressly specified in the design and technical specifications. If the erection risk is to be borne by the contractor, the contractor has to make sure that it has examined the project as approved by the competent authorities and has assessed that no other work, equipment, and service is needed to execute the project in addition to those included in the scope of works.

The contractor must oblige to a specific set of provisions and duties (*obligations of the contractor*). In particular, it must comply with the technical specifications and applicable law of the country of the project. It has the obligation to clear away and remove from the site any surplus materials and rubbish caused by the works according to the procedures and requirements set forth by the applicable law; at end of construction also all excess tools and machinery in order to provide to the client a clean and unencumbered site for correct and agile operations.

It must follow, comply with, and be responsible for all labor and social security, insurance, health and safety at work, and prevention of labor risks, where the design, construction, and operation of the Plant being also consistent with the same regulations; all consequences for injuries or damages resulting from the works to individuals or property will be borne the contractor who agrees to exempt and hold harmless the client from all responsibility deriving from non-compliance by the contractor during the term of the agreement of the above provisions of law.

The contractor agrees to disclose full information (*disclosure obligation*) and updates regarding the technical details of the supply and progress of the Works on a monthly basis and, at any time, upon written request of the client. Usually, a term of 3 days is giving to inform the client upon occurrence of any incidents, giving full details of the same, and specifying whether there might be any adverse impact on the project.

The contractor shall also agree to carry out at its own cost all tests and trials required by the applicable law to ensure the correct functioning of the project together with all tests related to project performance (*obligation to pass tests*).

It agrees as well to grant free entry to the client and its representative or advisors (*obligatory entry permit*) to the workshops, warehouses, and site at which the contractor is carrying out its scope of works activities, usually after having given prior notice (normally one day advance notice) and that the visit is carried out in working hours and does not affect the normal execution of the works.

Shall the need for *subcontracting* arise, specific obligation in subcontractors management are entered between the parties; in particular, the contractor may subcontract all or part of the works to one or more subcontractors without the need to obtain any consent from the client apart from the duty to inform the client of the identity and details of the subcontractor. Such assignment will bear the contractor responsible for the respect of the project technical specifications, the quality, the use of correct equipment, supplies or assemblies, the health and safety, and timing of all works carried out by the subcontractors as if they were carried out by the contractor itself. Depending on circumstances, the contractor can also be requested to maximize the use of local subcontractors to the extent that they can provide best commercially available terms and conditions with a know-how and quality level in line with the contractor standards.

Another specific contractor's obligation is the *duty of surveillance* for all equipment and materials located or stored at the site during the whole period of execution and tests of the project and normally until a defined date is reached where the tests are completed and construction is accepted by the customer.

In case an O&M contract for the project is entered between the parties(see Sec. 7.10), the contractor is obliged to comply to all the provisions set forth in such agreement. It is not at all unlikely that the firm that builds the asset, will be the same operating and maintaining it.

The *obligations of the client* articles serve the purpose to ensure that the client, for instance, proceeds with the right and duly payments to the contractor as accrued during the period of contractor's servicing of the plant and the client works in good faith alongside the contractor for the successful erection and completion of the project. Other client's obligations are to maintain valid the representations and warranties declared in the agreement and the respect of the mechanics of the contract, in particular the issuance of the "notice to proceed" (NTP) to the contractor as defined in the relevant articles.

On-site health and safety articles lay out the mechanics of how to comply with the H&S regulations during project construction. The procedures to be followed are highly dependent on each country' regulations and public body requirements. Usually, an H&S manager is requested to be assigned to the site, risk-analysis documents prepared and their recommendations strictly enforced on all people who is granted access to the site.

The contractor is requested to perform all obligations (*programming and execution of the works*) in the agreement in accordance with industry standards, applicable law, permits, and the project timetable. All works must be carried out with due care, diligence as can be expected of a fully skilled, experienced, and competent contractor.

Once the client has approved the final design, it notifies the contractor in writing and, usually within 5 business days from notice, the parties meet to proceed with the delivery of the relevant site. Upon delivery, the client sends to the contractor an official written *NTP* that kicks off the start of the works for that specific project. Usually, works are requested to start within 10 calendar days after delivery of the NTP. Another provision can be that the day after the date of delivery of the NTP, the client and the contractor meet and sign a *site takeover certificate* with attached a detailed description of the site actual conditions. Changes in the schedule can be agreed between client and contractor during an interim period of, usually, 6 months provided that previous approval has been given by the financial institutions in case of a financing agreement in place.

The contractor is responsible for the *transporting and storage* of all the equipment, goods and materials to the site and will also be responsible for all taxes, insurance, authorizations, labels, packing, delivering and unloading, bearing the costs and expenses in case damages or losses are incurred, discharge, reception, storage, and custody of the referred materials.

The contractor is responsible for the construction, assembly, and erection of the *civil works* necessary for full completion of the project. Prior to the start and acceptance of the corresponding NTP, the contractor must ensure that it has studied and inspected to its full satisfaction the site, all the available reports and studies (i.e., geotechnical, geomorphological, hydrogeological, archaeological, military...), has evaluated all the restraints and surroundings, accesses, and weather characteristics of the site. In this respect, the contractor has to make sure that it has examined and validated the characteristics of the project site and, eventually, the design of the structures will be adapted to the site without any increase of the price by reason of any of the aforementioned circumstances. In this respect, the contractor must ensure that, prior to the start of the works and acceptance of the corresponding NTP, the relevant site is valid, adequate, and sufficient for the performance of the projected works. Once the NTP has been accepted, it will be understood that the contractor has full knowledge of the site and the design to be carried out on such site.

The client retains the right to perform inspections of the works and equipment and the contractor must provide the client with full access to the site where the works are carried out, subject to compliance with current safety rules and policies. The right of inspection does not, of course, relieve the contractor from its responsibility for the proper performance of the works.

The contractor agrees to carry out the works within the terms specified under the timetable and complete the works by the completion date; the deadlines set forth in the agreement may only be changed due to the occurrence of the following circumstances: force majeure, suspension of the works by written order of the client (provided that the same is not motivated by previous action or omission of the contractor), written agreement of the Parties, and due to causes attributable to the client. Any eventual extension of the deadlines will be proportional to the delay introduced. If the contractor fails to meet the completion date, penalties will have to be enforced.

The parties nominate and vest their respective representatives with sufficient powers to enable them to take the obligations and exercise the rights, as applicable and to make all amendments, changes of orders, and notices deemed necessary for the best outcome of the project. Some of the representatives are, for instance, the works director, a professional assigned by the client to the project with the responsibility to ensure that construction is aligned with what represented in the EPC contract, to control compliance and quality of components, to monitor the site, to verify coordination between contractor and subcontractors, to perform some project-management task as well as to interface with the H&S manager to maximize respect of H&S regulations, and to prepare periodical reports on works progress.

From the date of commencement of the works and until the project is changed of ownership, the contractor takes the risk of loss and full responsibility for the cost of replacing or repairing any damage to the applicable works performed and all materials, components, equipment, supplies, and maintenance equipment (including temporary materials) that are purchased by the contractor for the permanent installation in or for use during construction. Of course, such responsibilities do not apply in case of negligence or misconduct of the client, who will bear the risk of loss and full responsibility starting from the date of signature of the certificates that mark the change of ownership of the project.

The contractor may suspend the works temporarily only after normally a couple of months from a written notification if the client fails to make any payment within, for instance, 60 business days of the date on which such payment is required to be made or if the client is in default of any of its obligations outlined in the previous relevant paragraph. In such case, the works will be reinitiated only once the client has duly remedied its defaults.[4] Suspension of works can also occur due to force majeure events or in case the client has not managed to reach financial close (or financing institution approval). If the financial close is not obtained within the following 2 months from the suspension date of the works, the contractor may terminate the contract. With respect to costs, the contractor shall examine the works affected by the suspension and proceed to cure any defect, deterioration, or loss occurred during the suspension period and the costs transferred to the Client. The timetable, including the completion date, will have to be postponed and adjusted for a number of days equal to the number of days of suspension.

Installation, commissioning and tests are the sole responsibility of the contractor and are carried out by the latter under the presence of the client. These tests are regulated under the *acceptance tests and change of ownership* section.

The certificates that mark the path towards the change of ownership are the following:

- The construction acceptance certificate
- The provisional acceptance certificate
- The final acceptance certificate (FAC)

The *construction acceptance certificate* (CAC) is issued after the successful *construction acceptance test*, which aims at verifying that the Plant has been built in compliance with applicable law, with the technical specifications and with the final design. The CAC is a statement that the Works are completed although it does not release the contractor from its obligations under the EPC contract in general and to comply with the norms applicable to the project type of installation in particular. If the client states that the works are not completed, it will deliver to the contractor a list with works to complete (a *punch list*) indicating a reasonable period to cure the defects.

Once the project is fully operational, namely when the construction acceptance tests have been successfully carried out and, therefore, a CAC has been signed by both parties and the plant has been running nonstop since a consecutive period of days to be determined between the parties and usually within a few days from CAC (not counting stopping periods caused by reasons not attributable to the contractor), then the *provisional acceptance test* (PAC) takes place. To these effects, the contractor has to notify the client in writing when the design, supply, construction, and installation of the plant are achieved. The client will have a time-frame (in the range of some weeks, normally not more than 2) to check and, in case of success, to issue the *provisional acceptance certificate* acknowledging the successful completion of the works. The procedures of how the test is performed is detailed in the annexes to the EPC contract. If defects exist that do not permit the full operational functioning of the plant, the client will not be obliged to sign the PAC and the contractor will have a period of, usually, one month, in which to make good the defects (as fully identified in a punch list separate document) and to carry out again the PAC. This term may be extended as a function of the size of the defect to be corrected. If the defects are minor and the asset is fully operational, the PAC will still have to be signed. All the risks assumed by the contractor until such time will be transferred to the client some calendar days after the contractor has so required to the client. As a

---

[4] In case financing is entered into with banks of financial institutions, a suspension event is regulated by a *direct agreement* with such institutions. More information on the direct agreement will be given in Sec. 7.9.

condition precedent to PAC signing and closing, the contractor normally must deliver to the client a set of documents like all as-built drawings, design, and guarantees for the main components used in the construction.

In case a plant fails to comply with the technical and/or mechanical warranties given by the contractor, the client will not be obliged for any reason to sign the PAC until such parameters have been fully met and if the contractor does not correct the defect within the agreed time-frames, the client will be entitled to reject the plant.

The *FAC* is issued after a longer period (in the range of 2 years) upon successful completion of the *final acceptance test* which entails that the asset has been performing according to agreed-upon performance and warranties (as described in details in the annexes to the EPC contract). Signature of the FAC may not be withheld, unless defects exist that have not been corrected and which do not permit the plant to be fully operative. Normally, the test is a verification of the performance after cross checking of the O&M reports described in the O&M agreement.

The *price and payment terms* section deals with the payment of the consideration for the project and the deadlines to proceed with such payments. Payments can be made according to different possibilities. A single, on-off downpayment for the total project consideration is usually unlikely, due to the many risks that can be incurred during the erection of the plant. This is why, especially during long construction times and complex projects, a set of milestones can be agreed upon between client and contractor, where installments and time schedules, together with sets of conditions precedent that constitute a prerequisite for money transfer, are negotiated between the parties. Some examples of milestones can be 20% of the total consideration at start of works, another 20% when main components arrive on site, other percentages at different other stages of plant construction or when after tests are passed, usually the final balance being paid at PAC. A particular way of proceeding with the payments is providing the contractor with a downpayment, say of 15–20% at start of works, with all the remaining 80--85% at PAC, or even proceeding with 100% of the total consideration at PAC; this way of paying the contractor is referred to as *vendor finance*, where is the contractor itself who completely bears all the burden of risk and its financing and waits for a lump sum payment when construction is ended successfully. The client, therefore, is financed by the contractor and doesn't need to resort to financial institutions to proceed with the project.

A condition precedent for money transfer is, for instance, that a draft of the invoice, before being sent to the client, shall be submitted to the technical consultant (usually assigned by financing institutions in case of financing in place) and to the works director, who shall confirm (within a period of time in the range of 7–10 business days), that the works have been progressing in line with the time schedule and that the milestone required for the issue of such invoice has been duly met by the contractor. The invoice can then be sent by the contractor to the client upon receipt of the above confirmation or in case the two professionals do not challenge the invoice draft within the same time period.

In case of delayed payments, the contractor can charge interest rates for the delay according to bank practice or agreed-upon interest rates.

One important provision is that the price can be determined on a fixed basis for the entire execution of the project. This means that the price will not be modified unless the parties agree upon any such change like, for instance, modifications or variations to the scope of works or to the technical specifications that will require price adjustments.

During the phases of the construction that move from signing of the contract, the CAC, the PAC, and the FAC, a set of guarantees must be put in place and managed to make sure the contractor performs accordingly to the provisions set out in the EPC contract (the *warranties and penalties*). After the client and the contractor have signed the EPC contract, at execution date a *completion guarantee* under the form of a first demand bank or insurance bond (from reputable institutions with a minimum S&P's rating) and guaranteeing the full and correct performance of the works by the contractor, must be delivered by the contractor to the client. Upon signature of the PAC, the client returns to the contractor the completion guarantee against the simultaneous delivery by the contractor to the client of a *performance guarantee* (usually for an amount equal to a percentage of the price) for the full and correct performance of all the undertakings assumed by the contractor specifically warranting that the design, the components and the assembly of the plant shall be free of defects for the period between PAC and FAC (the *warranty period*). Such guarantee shall be returned to the contractor once the FAC for the plant has been signed.

Apart from guaranteeing the correctness of the design, the contractor issues a warranty for defects which is the the obligation to repair or replace, if applicable, the materials and accessories that are ascertained to be defective, as well as in the labor, travel, and transportation used in the repair or replacement during the term of the warranty period. Usual request is to repair or replace any defective material or component within 30 business days from the ascertainment of the defect. In case the material or component of the affected plant does not work correctly and the contractor is not able to find out the reason, the contractor will be obliged to replace the material or component of the affected plant with a new one within 30 business days from the, say, third failed attempt.

Usually, a clause like the following is used to ensure that:

> ...nothing in this clause shall be intended or shall be interpreted so as to limit the Contractor's liability in case of fraud or gross negligence of the latter...

and

> ...the above penalty has been agreed by the parties with express exclusion of the right of the Client to claim for any additional or major damages.

The contractor will not be under any warranty obligations (*exclusion of warranties*) if damages have been caused or occurred due to:

- normal wear and tear
- the unwary use, operation and maintenance, repair, alterations not in compliance with the user manuals, technical specifications, or with any other instructions and guidelines of the equipment, components, and materials provided by the manufacturer or the contractor
- the use of materials, equipment, tools, accessories, services, schemes or designs not supplied, set up, or expressly authorized by the contractor
- force majeure events
- changes in legislation (the "change in law" event), subsequent to the date of execution of the PAC, imposing performance reductions, or changes in design
- any other causes or events which can be ascribed to the client

The *completion date of the works and penalties* section of the EPC contract takes care of what happens in case of delays in the timetable of the project. These provisions are meant to make sure that the contractor is applying the utmost diligence and care in reaching construction close in the correct amount of time. For the client, start of operations is the moment when it starts to receive the revenues from the asset underlying the project, therefore, every delay can make the client bear additional costs or revenue losses that must be compensated for in case the delays are attributable to the contractor. A maximum amount of penalties for delay are anyway normally limited to a percentage of the EPC price, although nothing in this clause is intended so as to limit the contractor's liability in case of its fraud or gross negligence.

All liquidated damages in an EPC contract (due to performance losses, delays...) are agreed not to exceed, in aggregate, a certain amount (for instance, the 20%) of the EPC price (the *maximum penalty cap*); in case at any time the maximum penalty cap is reached by the contractor, the client shall have the remedies specified under the following "Termination" section.

Termination clauses detail all the events leading to premature contract closing and its relevant consequences. Some of the events that lead to termination can be, for instance:

- a material breach
- the maximum penalty cap is reached
- a petition for winding up is instituted against the contractor by a court of competent jurisdiction
- any of the completion or performance guarantee becomes invalid or otherwise ceases to be effective
- expiry of bank guarantees before end of contract with the contractor failing to provide a replacement guarantee
- the contractor abandons the performance of the scope of work without reason for a period of more than 30 consecutive calendar days or ninety calendar days in any annual period
- any serious and repeated default of the contractor that impedes or materially affects the successful performance of the contract
- the contractor not maintaining valid insurances
- the contractor assigning the agreement or subcontracting the execution of the works in breach of the limitations and restrictions specified under the contract
- the contractor becomes bankrupt or insolvent, is in liquidation, has a receiving or administration order made against it, enters into a creditors arrangement or carries on business under a receiver, trustee, or manager for the benefit of his creditors, or if any act is done or event occurs which (under applicable law) has a similar effect to any of these acts or events, unless the client consents to the contract continuing to be effective

In case any party commits a material breach of its obligations, the other party has the right to notify the defaulting party to remedy the breach in a term of not less than usually 30 days, with termination if the remedy actions are not performed by the defaulting party.

The *effects of termination* depend if the termination is requested by the client or by the contractor. If the termination is performed by the client for causes attributable to the contractor, the former can choose one option between the following set of actions:

- retain the ownership of the equipment, the materials, the works, the services received by the contractor until the date of notification of termination, retain the right to be promptly returned any amount paid in relation to works or services that have not been concluded yet, require the contractor to bear all necessary breakage costs in connection to the financing arranged by the client and be paid as loss of profit an amount equal to a percentage of, usually, 15% of the EPC price.

- retain the ownership as in the previous clause and have the right to an indemnity for the proven damages to be decided in court.

- subject to the prior written consent of the financing entities (if the project has been financed by them), sell to the contractor the SPV owning the project at a price fixed by an external consultant or by an International Chamber of Commerce (ICC) chosen by the parties. Usually, the contractor shall bear any costs and expenses for any consultant service.

- the contractor shall retain the ownership of all the equipment and leave the project site in the same condition as it was before the start of works and repay all debt drawn down by the financing entities until the date of notification for termination including all necessary breakage costs in connection to the financing arranged by the client for the project, return to the client any amounts paid and pay as loss of profit an amount equal to, usually, 10% of the EPC price.

If the termination is requested by the contractor for causes attributable to the client, the former can choose one option between the following set of actions:

- the client retains the ownership of the equipment, the materials, the works, and the services received until the date of notification for termination of the agreement, but will pay the contractor for all of the previous until the date of notification of termination, for all the amounts spent for the equipment already ordered and for the loss of profit usually for a percentage set at 10% of the EPC price.

- subject to the prior written consent of the eventual financing entities (if the project has been financed by them), the contractor is entitled to purchase the SPV owning the project at a price fixed by an external consultant or by an ICC chosen by the parties. Usually, the client shall bear any costs and expenses for any consultant service.

The contractor's right to exercise the option to terminate the EPC agreement is subject to the contractor complying to the provisions set forth in the direct agreement that will be discussed later in this paragraph.

To protect the investment, a full section on *insurance* provisions regulates the contractor's and client's insurances that must be put in place to protect the EPC contract. Usually, contractor's insurances are required to be of primary standing (insurance companies with ratings not lower than S&P's A ratings):

- general third party civil liability insurance
- general own civil liability insurance
- marine and land transit
- EARs and advanced loss of profit
- insurance against accidents at work
- insurance against civil liability for the operation of vehicles and machinery
- any other compulsory insurance according to laws in force

The contractor must procure that all its subcontractors will enter into the same insurance policies mentioned above based on their involvement in the performance and supply of the Services so that the maximum indemnity coverage covered by the such insurance policies and those entered into by the contractor is not lower than the minimum thresholds specified in the EPC agreement.

The contractor is requested to procure a waiver of subrogation, so that the client and eventual financing parties shall be released from any responsibilities and the right to ask the two parties for compensation be waived, both by the insured party and by the insurance company, in connection with the risks covered by the signed insurance policies.

In case of changes incurred by the contractor that can affect the validity or the conditions of the insurance policies during the EPC period, the contractor must notify the client, whereas any breach or not keeping in full force and validity the insurance policies can led to the client independently obtaining an equivalent coverage (and debiting the contractor for all incurred expenses) or terminate the contract. Copies of the undersigned policies and the receipts of paid premia must be provided to the client usually within a 10-day period following the effective date.

Usually, all deductibles, losses, or damages in excess of the insured limits in the insurance policies are at the risk of the contractor.

Neither party may proceed with the *assignment* of the EPC contract, not even any rights or obligations, to any third parties without the prior express written consent of the other party. In partial derogation, a party (the *assignor*) may decide to assign to a third party (the *assignee*) only if the assignee is a company belonging to the same group of the assignor; this can occur only if communicated by written notice in a short period of time (usually 10 business days) in advance of the date of the proposed assignment, providing full information about the name and address of the assignee, relevant payment details, the level of the technical and financial capacity, which shall be the same of the Assignor. The assignor will remain jointly and severally responsible with the assignee. The client (the *assigned party*) can deny the assignment only if the assignee company has not the same level of technical and financial capacity of the Assignor. Credits arising under the EPC contract can be assigned to third parties like banking or financial institutions. If the client wants to enter into a project finance contract with banks or financial institutions, the contractor is required to cooperate with the client and avail the requests of the financing parties.

If the contractor needs to subcontract part of the works, he is allowed to do so with no necessary consent by the client apart from having the obligation to inform the client in writing, usually within a week and of course remaining fully liable for the actions carried out by the subcontractors, whose services and used components must comply with the provisions of the EPC agreement.

Force majeure events cannot be imputed to negligence or fraud of the affected party and cannot not be avoided or limited using ordinary diligence and result in damages or delays. Examples of force majeure events can be:

- adverse meteorological conditions like drought, hail, heavy or torrential rain floods, tornadoes, fires, land slides, lightening strikes, or other adverse natural phenomena, which prevent the contractor to perform the works

- epidemics

- labor strikes, not internal to one of the parties' businesses

- war, civil conflicts, acts of sabotage, terrorism, vandalism, and embargoes

- changes in applicable legislation, the revocation or suspension of any authorisation, permit, or license, or any other decision or act of any authority, which cannot be ascribed to one party

- climate conditions that exceed those for which the plant was designed and that are detailed in the respective technical specifications of such plant

- climate or meteorological conditions that, according to health and safety laws and regulations, make the access to a site, and/or the execution of the works unsafe or nonviable

The party affected by a force majeure event has to promptly notify the other party and both parties have to perform best efforts and cooperate in good faith to mitigate and limit, to the extent possible, the effects of such event. The lead times and delivery dates and plans specified under the EPC contract shall be adjusted accordingly to the extent of the force majeure event, with each party bearing its own costs. In case of events lasting more than an agreed-upon period of time (i.e., 180 consecutive calendar days, or more than 210 nonconsecutive days), the affected party can retain the right to withdraw from the contract.

The *transfer of title and risk* section deals with transferring of the ownership and henceforth the risk, of the asset. Specifically, the actual transfer of ownership of any equipment to be delivered by the contractor occurs on the date on which the delivery of the equipment is made at the contractor's premises, with its obligations to custody, mark such equipment as property of client and within a certain number of days of receipt, to send to the client a document containing all the identification details for client's proper accounting of the equipment. The transfer of the plant will be made, for all contractual purposes, upon the issuance of the PAC and the client will bear the risk of loss and full responsibility for the asset.

From the date of commencement of the works as specified in the NTP and until the signature of the PAC, the contractor takes the risk of loss and full responsibility in relation to all the equipment and the plant. The contractor bears the costs of replacing or repairing any damage to the applicable works performed and all materials, components, equipment, supplies, and maintenance equipment that are purchased by the contractor for the permanent installation in or for use during construction, even if of the property of the client. Of course, the contractor is not held liable in case of client's misconduct causing damages or losses to the equipment and the plant.

The R&W section is an important part of EPC contracts and share the main characteristics described under Sec. 7.5. Some typical EPC representations and warranties entail that the contractor can fully and lawfully dispose of the site of the project and has obtained

> ...any and all authorizations or consents of any authority or third parties which are required for the purposes of the services to be performed and that the Contractor is acquainted to all local, regional, and national conditions, which could affect the scope of the Agreement, accepting all responsibility for having properly evaluated as at the date of execution all costs and contingencies for successfully performing the services and satisfying all obligations and agrees to bear all and any consequences resulting from any improper evaluation.

Finally, paragraphs on *confidentiality, applicable law, arbitration*, and *miscellaneous* complete the contract structure. The confidentiality section deals with what we have seen under paragraph 7.6, the applicable law article represent under which country's law system the EPC contract is enforced, the arbitration is the process of amicably settle disputes before taking them to court if the parties cannot convene on an amicable settlement alone, usually by assigning arbitrators and the rules (normally of the local chambers of commerce) of arbitration that will take a decision by majority or unanimous approval within a period of time in the range of 4–6 months from appointment of the arbitrators. Miscellaneous agreements are then outlined to provide rules on, for instance, the nondisclosure of the deal entered by the parties with the EPC contract, the payment of expenses, and taxes in relation to the EPC agreement. The *severability* clause maintains the validity of the remaining contract provisions in case some of the provisions are declared invalid or are unenforceable. In such case:

> ...the Parties shall be released from all rights and obligations which derive from the provision declared void, invalid, or unenforceable, but only to the extent that these rights and obligations are directly affected by such voidness, invalidity or unenforceability. In this event, the Parties shall negotiate in good faith to substitute the void or invalid provision with another valid and effective provision which sets out, to the extent possible, the original intention of the Parties.

Annexes are attached to the EPC contract main body to provide complementary documents such as the legal authorization from local administration for the construction of the project; the final design and technical specifications; list of equipment needed for construction; product certifications and guarantees; time-table of works and milestones; completion and performance guarantees; acceptance tests procedures; health and safety documents; and insurance policies. To protect the client against contractor's underperformance, a guarantee can be asked by the client in the form of bonds or sureties from primary banks or insurance companies.

When a project is, or can be, financed by third parties, typically a *direct agreement* letter is attached as draft or executed copy[5] that binds the O&M contractor with the financial parties that are monetarily sponsoring the project in order to assure the financial parties that the project operations will continue smoothly even in the event of a default of the organization that has entered into the credit agreement with the financial institution by borrowing the capital needed to implement its business idea.

---

[5] An *executed* copy is a document signed by all involved parties and showing the date in which that signing has been performed.

The main entities in a direct agreement are the borrower; the contractor; the *finance parties* that can be banks or an institution acting as an *agent* on behalf of the banks, its affiliates or subsidiaries; or an *eligible person* who can be the agent, the bank or one of its subsidiaries, or any other person that can be approved by the contractor, provided that such approval be not unreasonably withheld or delayed and such eventual withholding be based on legal capacity, power, authorization, technical ability to operate and maintain the project, its financial standing or the financial standing of any guarantor, or, finally, its position as a direct competitor in the market or directly controlled[6] by a direct competitor.

Under the direct agreement, the contractor agrees in particular to behave following a specific set of rule. Between them, until and unless it receives notice to the contrary from the agent, all amounts payable to the borrower under the agreement or the bonds shall be paid to the bank accounts that are under control of the financing parties; for no reason, terminate or suspend performance of its obligations under the agreement prior to the expiry of the term of at least a certain number of weeks, usually in the range of a 2–3 months and informing the agent about the proposed date of termination, the reason for termination, the nature, and amount owed by the borrower to the contractor (which must be compiled with care and in good faith).

The contractor might anyway suspend the performance of its obligations under the agreement (but not terminate them) after a period of usually a week from the day of receipt by the agent of the termination notice in the event of nonpayment of an invoice or in case of a force majeure event as outlined in the relevant article of the contract.

To ensure continuity of operations, the agent can act on behalf of the borrower during the suspension of contractor's services and ask the contractor to continue performing its services subject to payment of the related price as negotiated in the contract.

The agent may also procure that an eligible person assume, jointly and severally[7] with the borrower, all the borrower's rights and obligations identified in the notice in the case of a termination event and those which arise after the date of such undertaking (the *additional obligor*). The contractor will automatically approve the additional obligor if he is one of the finance parties.

The direct agreement defines a *step-in period*, which represents the time frame during which the additional obligor takes charge on behalf of the borrower; a *step-out date*, when the additional obligor is released, upon preliminary notice from the agent to the contractor (usually a couple of weeks) from any obligations to the contractor except for those amounts who have not been paid and were invoiced before or during the step-in period. During the step-in period, the agent may, upon usually a 30-day prior written notice to the contractor and any additional obligor, procure the assignment of the borrower's rights and liabilities of the agreement to another eligible person (the *novation*[8]) provided that the same amounts in the previous paragraph have been paid and as a consequence the additional obligor's rights to the contractor will thereafter cease to be in force.

---

[6] Usually, *control* is intended as the power to direct the management and the policies of an entity, whether it be, for instance, by contract or ownership of voting rights.

[7] A *joint and several liability* occurs when a claimant can seek for remedies against two defendants as if they were a single liable party, while it's in the the responsibility of the defendants to decide about their respective proportions of liability and payment.

[8] In business law, the *novation* is the replacement of an obligation to perform or a party with a different obligation or party and is valid only with the consent of all parties that entered the original agreement.

As an additional clause in the direct agreement, the contractor is not allowed to transfer or dispose of its rights and obligations unless the agent so agrees, if that the new legal person is technically and economically capable of carrying out the tasks defined in the contract and enters in the DA as the new contractor. Finally, the contractor is prevented, at least for a period in the range of 3 months, from filing for any insolvency proceedings against the borrower or any of its assets.

A structure of a typical EPC contract is given in Table 7.2.

7.10   Operations and Maintenance Contract

An *operations and maintenance* (O&M) contract is an agreement, entered into between the client and the O&M provider (or contractor), that defines all activities and provisions binding the two parties to ensure that the energy plant achieves the agreed-upon performances and is capable of maintaining such performances over the contracted period of time.

The contract starts as usual by stating the execution date and providing the *by and between*, the *whereas*, the *recitals and definitions* and the *terms of the agreement* sections, where this latter paragraph defines the duration of the validity of the contract and its eventual automatic renewal after a certain number of years.

The *scope of works* and *obligations of the contractor* paragraphs state that the client entrusts the contractor with the provision of the services outlined in the paragraph and gives a list of all the services required from the contractor and becoming part of the obligations of the contractor: preventive and corrective maintenance of the assets, the management of its operations, and the monitoring and surveillance, reporting. All services are described with precise reference to specific performance needs peculiar to that kind of service. For instance, for scheduled and preventive a minimum number of periodical inspections are carried on the site, that all electrical, structural, and civil works controls, which are compulsory by law are respected with the correct timing and procedures; in case of failure or malfunction, the contractor is requested to carry out an error analysis and undertake all the necessary steps to repair, test, and restore full functionality of the plant; depending on type of failure or malfunction, the requested intervention times can be requested to start immediately, in no less than for instance 24 h from detection or 72 h in case of malfunction not affection production. Spare parts should be kept in the right quantity and storing conditions in warehouses not farther away than a defined distance from the plants and used parts to be replenished not later than a period of time in new or fully functioning refurbished conditions. System security can be requested to be performed 24 h a day for 365 days of the year, either by remote surveillance or patrolling, with on site intervention in a case of intruder alarm set off. Reporting should be performed, with correct methodology and predefined formats, after scheduled site visits for normal inspections or corrective maintenance interventions. Usually, annexes attached to the O&M contract are meant to outline with completeness of details all the requested services.

Provisions can indicate that costs that must be borne for labor, repair, replacement of parts for corrective maintenance services be included in the price due to the contractor, while the purchase, package, delivery, and transportation of parts needed for repairs will be borne by the client. An article about the *exclusion of liability* protects the contractor

Main body:	
	By and between
	Preamble ("whereas")
	Recitals and definitions
	Terms of the agreement
	Scope of works
	Changes and exclusions to scope of works
	Obligations of the contractor
	Obligations of the client
	On-site health and safety
	Programming and execution of the works
	Acceptance tests and change of ownership
	Price and payment terms
	Warranties and penalties
	Completion date of the works and penalties
	Termination
	Insurance
	Assignment
	Sub-contractors
	Force majeure
	Transfer of title and risk
	Representations and warranties
	Confidentiality
	Applicable law
	Arbitration
	Miscellaneous
Annexes:	
	Project construction authorization certificate
	Technical specifications, final design, and site layout
	List of equipment
	Acceptance tests
	Completion guarantee
	Performance guarantee
	Project timetable
	Product certifications and guarantees
	Performance evaluation calculation
	Ordinary maintenance programme
	Health and safety documents
	Direct agreement

TABLE 7.2 Typical Structure of an EPC Contract

from the obligation of providing the services if the client has performed interventions, modifications, or used parts and components without respecting the correct technical specifications provided by the manufacturer of the plant or by the contractor and not expressly authorized by the contractor in writing, unless the client intervention has not been made necessary by misconduct of the contractor itself. An important obligation for the contractor is to be in compliance with and to undertake all necessary activities to ensure health and safety of the premises and its operating procedures in accordance with applicable health and safety laws and regulations.

Another paragraph covers the *price and terms of payments*, where payments are usually due in periodic installments made in arrears and subject to interest charges if the client delays the payments.

A paragraph is devoted to define the *obligations of the client* to ensure that the client, for instance, proceeds with the right and duly payments to the contractor as accrued during the period of contractor's servicing of the plant and the client works in good faith alongside the contractor for the successful operations and maintenance activities.

The performances of the project, together with eventual over or under performance monetary compensations, are regulated under the *performance of the project* section. Usually, performance calculations are included in the annexes to the contract, since they can entail the use of complex math and formula definitions that can be better encapsulated in stand-alone documents in order to facilitate potential modifications to the performance calculation methods without affecting the already complicated main body of the O&M contract. In case of verified over performance of the contractor that increases operational results of the plant over some negotiated target performance ratios or thresholds, the contractor can be entitled to receive a bonus over the contract price that is usually a portion of the benefits accrued to the client as a result of contractor's activities. On the contrary, if the contractor is not meeting the guaranteed performances, will be subject to compensation to ensure that client's business profitability goals are respected. The mechanics of these *bonus and malus* structures are depending on the revenues model that is peculiar to each project and are normally defined on a case-by-case basis. Furthermore, the parties usually agree to a maximum amount of yearly liquidated damages from the contractor to the client as result of the failure to reach the guaranteed performance; to give an idea of the numbers, a potential maximum for liquidated damages shall not exceed 150–200% of the negotiated O&M price. In case such maximum liquidated damages are incurred, the client can have the option to terminate the contract with the contractor.

If the performance of the plant is affected by *force majeure* events, there are contractual provisions in place that cancel contractor's liability when such performance reductions cannot be ascribed to its bad plant management. Such events not only can be of the like of extreme meteorological conditions, acts of sabotage, upheavals, terrorism, vandalism, but also changes in current laws and regulations or revocations of the client's rights on lawfully own and operate the project, up to seizures or forfeitures, land confiscation, or expropriation. In case a force majeure event happens, only the portion of the plant that is affected is subject to be out of contractor's performance obligations. Then, if such an event should last for more than a certain period of time, usually in the range of some months, the contractor can withdraw from the contract, usually by written communication and without prejudice to the rights or remedies that have been accrued during the period until withdrawal effective date.

To protect the investment, a full section on *insurance* provisions regulates the contractor's and client's insurances that must be put in place to protect the O&M contract.

Usually, contractor's insurances are required to be of primary standing (insurance companies with ratings not lower than S&P's A ratings):

- general third-party civil liability insurance
- general own civil liability insurance
- insurance against civil liability for the operation of vehicles and machinery
- any other compulsory insurance according to laws in force

The contractor must procure that all its subcontractors will enter into the same insurance policies mentioned above based on their involvement in the performance and supply of the Services so that the maximum indemnity coverage covered by the such insurance policies and those entered into by the contractor is not lower than the minimum thresholds specified in the O&M agreement.

The contractor is requested to procure a waiver of subrogation, so that the client and eventual financing parties shall be released from any responsibilities and the right to ask the two parties for compensation be waived, both by the insured party and by the insurance company, in connection with the risks covered by the signed insurance policies.

In case of changes incurred by the contractor that can affect the validity or the conditions of the insurance policies during the O&M period, the contractor must notify the client, whereas any breach or not keeping in full force and validity the insurance policies can led to the client independently obtaining an equivalent coverage (and debiting the contractor for all incurred expenses) or terminate the contract. Copies of the undersigned policies and the receipts of paid premia must be provided to the client usually within a 10-day period following the effective date.

Usually, all deductibles, losses, or damages in excess of the insured limits in the insurance policies are at the risk of the contractor.

The client will be willing to contract:

- all risks property and business interruption insurance
- third-party liability insurance

with a minimum set of terms like:

- an endorsement of the client, the contractor (and its subcontractors), and eventual other parties (i.e., financing parties) indicated as coinsured
- a cross liability clause
- a loss payee clause on behalf of eventual financing parties

To maintain symmetry between client and contractor, the client can be requested to sign a waiver of subrogation, although it can be safely avoided if the contractor does not expressly request the waiver. To complete the clauses about insurance policies, any cancellation or material modification of the policies must be in writing at least, usually, 30 days in advance.

Neither party may proceed with the assignment of the O&M contract, not even any rights or obligations, to any third parties without the prior express written consent of the other party. In partial derogation, a party (the assignor) may decide to assign to a third party (the assignee) only if the assignee is a company belonging to the same group of the assignor; this can occur only if communicated by written notice in a short period of time (usually 10 business days) in advance of the date of the proposed assignment, providing full information about the name and address of the assignee, relevant payment details, and the level of the technical and financial capacity which shall be the same of the assignor. The assignor will remain jointly and severally responsible with the assignee. The client (the *assigned party*) can deny the assignment only if the assignee company has not the same level of technical and financial capacity of the Assignor. Credits arising under the O&M contract can be assigned to third parties like banking or financial institutions. If the client wants to enter into a project finance contract with banks or financial institutions, the contractor is required to cooperate with the client and avail the requests of the financing parties.

If the contractor needs to subcontract part of the works, he is allowed to do so according to the *subcontract* section of the agreement. Subcontracting needs no necessary consent by the client apart from having the obligation to inform the client in writing, usually within a week and of course remaining fully liable for the actions carried out by the subcontractors, whose services and used components must comply with the provisions of the O&M agreement.

A section on the *suspension of services*, protects the contractor by granting it the right to suspend its activities, on a temporary basis and after having noticed to the client at least 1 month in advance, if the client doesn't pay the contractor after the terms negotiated in the agreement. Such activities will be restarted after the client has paid what due. During suspension, the contractor is requested to use all reasonable efforts to minimize costs arising during the period, while being suspended from its obligation to achieve the performances of the project. Also the client can request the contractor to suspend some services, but keeps the contractor free from obligations on guaranteed performance for the period of time when the contractor is asked to suspend.

Under the articles in the *termination* section, both parties have the right to terminate the contract following a material breach by the other party, granting to the defaulting party a term of not less than usually 1 month to remedy the breach. The client will have the right to terminate the contract if the contractor reaches the maximum limit in liquidated damages or causes the project to have low performance ratios or low availability (like, for instance, 4–6% under the 98% plant availability); being in default of setting up or keeping in force the necessary insurance policies and not remedying such default within a certain number of days (i.e., 2 weeks) of the receipt of a written notice; breaching the assignment and sub-contractor clauses; interrupting without justification the services for a period longer than a negotiated length of time (2 weeks being an industry standard); going bankrupt. On its part, the contractor will have the right to terminate the contract if the client does not comply with its obligation to pay the contractor unless the nonpayment is remedied within a certain number of days (i.e., 20–30) from the due date. When the termination arises due to a default of the contractor, the latter has to remove from the project site, bearing all costs, every machinery, parts, and components that have not been already incorporated in the project as per the O&M contract provisions.

In some cases, the client can request the contractor to issue a *guarantee* like sureties or bonds from primary banks or insurance institutions. This guarantee can be enforced in case the contractor does not fulfill its obligations.

The R&W section is, as always, an important part of O&M contracts and share the main characteristics described under Sec. 7.5. Some typical O&M representations and warranties entail that the contractor can fully and lawfully dispose of the site of the project and has obtained

> ...any and all authorizations or consents of any authority or third parties which are required for the purposes of the services to be performed and that the contractor is acquainted to all local, regional, and national conditions which could affect the scope of the agreement, accepting all responsibility for having properly evaluated as at the date of execution all costs and contingencies for successfully performing the services and satisfying all obligations and agrees to bear all and any consequences resulting from any improper evaluation.

As in other contracts like the EPC in Sec. 7.9, paragraphs on *confidentiality, applicable law, arbitration,* and *miscellaneous* complete the contract structure. The confidentiality section deals with what we have seen under Sec. 7.6, the applicable law article represent under which country's law system the O&M contract is enforced, the arbitration is the process of amicably settle disputes before taking them to court if the parties cannot convene on an amicable settlement alone, usually by assigning arbitrators and the rules (normally of the local chambers of commerce) of arbitration that will take a decision by majority or unanimous approval within a period of time in the range of 4–6 months from appointment of the arbitrators. Miscellaneous agreements are then outlined to provide rules on, for instance, the nondisclosure of the deal entered by the parties with the O&M contract, the payment of expenses and taxes in relation to the O&M agreement. The *severability* clause maintains the validity of the remaining contract provisions in case some of the provisions are declared invalid or are unenforceable. In such case:

> ...the parties shall be released from all rights and obligations which derive from the provision declared void, invalid, or unenforceable, but only to the extent that these rights and obligations are directly affected by such voidness, invalidity, or unenforceability. In this event, the parties shall negotiate in good faith to substitute the void or invalid provision with another valid and effective provision which sets out, to the extent possible, the original intention of the parties.

Annexes are attached to the O&M contract main body to provide complementary documents such as the: legal authorization from local administration for the construction of the project; the as-built design; list of spare parts needed for maintenance of the project; product certifications and guarantees; scheduled activities over 1 or more years with detailed description of such activities and their timing; details and mechanics of calculation of performance ratios and bonus or malus that the contractor con be entitled to receive in case of good performance, or pay back in case of bad performance, plant availability, performance decays; ancillary systems (i.e., monitoring, antitheft, and anti-intrusion); health and safety documents; insurance policies; and sureties or guarantees to protect from contractor's under-performance. To protect the customer against contractor's underperformance, a guarantee can be asked by the client in the form of bonds or sureties from primary banks or insurance companies.

When a project is, or can be, financed by third parties, typically a direct agreement is attached as draft or executed copy. Details of a direct agreement have already been highlighted in Sec. 7.9.

An example of structure for a typical O&M contract is given in Table 7.3.

Main body:	
	By and between
	Preamble ("whereas")
	Recitals and definitions
	Terms of the agreement
	Scope of works
	Obligations of the contractor
	Obligations of the client
	Price and payment terms
	Performance of the project
	Force majeure
	Insurance
	Assignment
	Subcontractors
	Suspension of services
	Termination
	Guarantees
	Representations and warranties
	Confidentiality
	Applicable law
	Arbitration
	Miscellaneous
Annexes:	
	Project construction authorization certificate
	As-built design
	Spare parts list
	Product certifications and guarantees
	Ordinary maintenance programme
	Performance evaluation calculation
	Monitoring and reporting
	Anti-intrusion and anti-theft systems
	Health and safety documents
	All-risk insurance policy
	O&M guarantees
	Direct agreement

TABLE 7.3 Typical Structure of an O&M Contract

7.11   Joint Venture Agreement

The *joint venture agreement* (JVA) can be one of the natural consequence of a successful PDA. A JVA structure is similar to all other contracts or agreements. It begins with the usual introduction of the execution date, the parties, the whereas, and the purpose of the agreement. The recitals and definitions complete the introductory phase of the JVA by defining the eventual "brand" or "trade marks" of the machinery or systems in case of a commercial agreement, "intellectual property rights (IPRs)," "minimum quantity," "net sale prices," "order quantity," "reserved territories," "assigned territory," mechanics of the interpretation of particular words like "hereof," "herein," "including," "in particular," "for example."

The *appointment* set of articles can be introduced to state that one party is appointed by the conceding party as distributor, marketer, or responsible of the aim or purpose of the agreement in the assigned territory. For instance, if one party is appointed as the sole marketer of a system, the conceding party can request that the systems be exclusively supplied to the marketer who will only buy such systems from the conceding party. Other provisions apply for sharing the know-how and protecting the *IPRs* from the conceding party to the other party, eventually even on the *variations* that can be devised and performed by one of the parties. The *title and risk* clause regulates the transfer of risk and property of the asset. Normally, risk is transferred when the asset is collected, while the ownership of the asset does not pass until payment is received.

Other provisions on *quality control* and *marketing and sales* are circumstance-specific, depending on the type of the relationship between the parties, the product or service, and the countries of interest. Most of the remaining provisions are common in other types of contracts and have been described in details in the previous paragraphs. An example of a JVA structure is given in Table 7.4.

7.12   Quota Purchase Agreement

A *quota purchase agreement* (QPA), or a *share purchase agreement* (SPA), is the contract between two or more parties that states the terms and conditions for the acquisition of a portion or the total of the corporate capital of a firm. In energy projects, such firms can be the SPV that owns the energy assets, the ownership of which is transferred from one party to another where, usually, one party is the development and construction company and the other party is the investor and future asset manager of such assets. A QPA is normally the outcome of the process that started with an NDA, followed by an initial LOI that led to a full DD analysis of the project, after which the investor has decided to continue with the asset acquisition.

Some important provisions in the QPA are related to the timing that leads the acquisition process to the *closing*, the fulfillment of all the activities to be carried out by the parties on the *closing date*, which is normally the first business day that follows the date on which the conditions precedent defined in the QPA have been satisfied or waived. The *reference date* is the day at which the transaction price has been determined but on which the control of the firm is not yet on the acquiring party side. The *execution date* is the day in which the parties convene and sign the QPA. The *long stop date* is the last day of the period during which the conditions precedent must be met or waived for the agreement to proceed and avoid termination based on CP.

Main body:	
	By and between
	Preamble
	Recitals and definitions
	Appointment
	Provision of know-how
	Quality control
	Marketing and sales
	Confidentiality
	Sale of product
	Title and risk
	Protection of IPRs
	Liability, indemnity and insurance
	Obligations of the parties
	Subcontracting
	Assignment
	Duration and termination
	Waiver
	Entire agreement
	Variations
	Severance
	Third party rights
	Force majeure
	Notices
	Dispute resolution
Annexes:	
	Product technical specifications
	Product guarantees
	Claim management
	Product performance
	Price list

TABLE 7.4 Example of Structure of a JVA Agreement

As a way to cater for adjustments in the transaction price, an analysis of the final accounts at the reference date is performed to establish the *NFP*, which is the difference between total short-term assets and short-term liabilities. The assets may include:

- trade receivables, including invoices issued and to be issued
- advances paid to suppliers
- accrued revenues and prepaid expenses
- VAT balance

- tax receivables
- cash and cash equivalents

and the liabilities may include:

- trade payables, including invoice received and to be received
- costs and expenses in the interest of the firm
- accrued liabilities and deferred expenses
- tax payables

All the costs and liabilities for the accomplishment of all conditions precedent are usually considered between the liabilities, barred different agreement between the parties. As a result, the NFP is a positive number equal to the amount of the excess of the aggregate assets over the aggregate liabilities, or a negative number equal to the amount of the excess of such liabilities over such assets.

The period of time between the reference date and the closing date is the interim period during which many events can occur in the life of the firm and must be regulated by the provisions in the QPA.

The *permitted leakage* is the definition of all the expenses incurred by the firm to be acquired during the interim period consisting and needed to preserve the value of the assets and the ordinary course of business of the SPV owning them. These expenses entail, for example, the proquota fees due to payments to contractors, land surface rights, public authorities, and insurance premia.

All expenses occurred between the reference date and the execution date are subject to revision by the acquiring party, usually within some days after the execution date and, in case of disagreement on the qualification of such the expenses as permitted leakage, there will be a recourse to arbitration. With regard to the period from the execution date and the closing date, normally a monthly threshold is granted to the party operating the firm. For expenses over the monthly threshold, there must be approval by the acquiring party, which may not be unreasonably withheld and shall be deemed granted if no written reply is received by the seller within, say, two business days from the date of receipt. If the purchaser denies its approval, the seller may decide whether not to carry on such expenses or proceed but with recourse to arbitration.

The agreement structure is the same of the contracts described in the previous sections. The contract starts with the usual "by and between", "recitals" and "definitions", where all the concepts previously discussed are outlined and defined.

The *transaction* article is the irrevocable promise that constitutes one of the main provision for the contract to be effective, as discussed in Sec. 7.3:

> Subject to the Conditions Precedent set forth in article (. . .), the Seller irrevocably undertakes to sell and transfer to the Purchaser, and the Purchaser irrevocably undertakes to purchase from the Seller, on the Closing Date, the Participation and to fulfil all their respective obligations in accordance with the terms and conditions set out in this Agreement.

The conditions precedents are listed in the article following the transaction article. They depend upon the peculiarities of the asset acquisition and if not met or waived, they give rise to the right of the acquiring party to terminate or withdraw from the QPA.

The parties agree in the *consideration* article about the transaction price which is subject to an increase or a decrease on the basis of the NFP and the permitted leakage and the final accounts. Since the transaction price is determined on a cash and debt-free basis, the NFP is completed normally within a month after execution date.

The *closing and actions at closing* section details all the activities to be performed at closing; normal actions to be taken are the delivery of the resignation letter of the SPV directors with effect from the closing date, including a formal waiver of any and all claims that such directors may have against the SPV until the closing date, the SPV corporate and accounting books, a written statement attesting that the permitted leakage has not changed and that the participation is free from any encumbrance[9] at the closing date, the payment of the transaction price. When all provisions in the QPA are met, the *deed of transfer* is the notary agreement that the parties execute on the closing date in order to make effective the transfer of the participation from the selling party to the purchasing party. Then, after a period in the range of two months following the closing date, the acquiring party is entitled to request the repayment of all costs, expenses, debts, or liabilities not included in the permitted leakage, the NFP, and the final accounts. All the activities that take place during the closing are considered as being a single, or *sole*, legal transaction, so that no action is deemed to have taken place if and until all other actions and transactions constituting the closing have taken place.

The usual "Representation and Warranties" section follows and is represented for both parties, the seller and the purchaser. A set of articles are then introduced to provide the purchaser with the "indemnities" by the seller, who:

> . . . undertakes to indemnify the purchaser against all and any actual losses, damages, liabilities, and costs, with the exclusion of financial loss and loss of profits incurred by the purchaser as a consequence of any breach, inaccuracy, or untruthfulness of the representations and warranties given by the seller under this agreement; any other breach of the seller's obligations and undertakings set out under this agreement.

Depending on the type of R&W to protect against breach, there can be a time-bar after which such breach is no longer enforceable. Furthermore, the acquiring party shall not be entitled to claim any indemnification if any loss does not exceed a certain amount of money (called the *de minimis* clause), but in case the "de minimis" is exceeded, the indemnification will be enforced for the entire amount of the loss and not only for the amount which exceeds the "de minimis". Except in case of the seller's fraud or gross negligence, the seller's aggregate cumulative liability for any and all claims of the Purchaser or any third party under and in connection with this agreement can be limited to a cap (i.e., 20% of the Transaction Price) which cannot be applicable in some exceptions defined in the agreement itself. No breach or inaccuracy of any seller's R&W gives the purchaser the right to withdraw from the QPA unless if the amount exceeds an agreed-upon threshold. Finally, to protect the purchaser, a guarantee can be requested to the seller or its guarantor.

Ending the QPA is the usual set of articles related to confidentiality, applicable law, arbitration and miscellaneous provisions. Between the Annexes, the *power of attorney* is the document that gives to representatives of the seller and purchaser parties the

---

[9] An *encumbrance* is a burden like a mortgage, pledge, option, foreclosure, seizure, confiscation, security interest, that are enforced on an asset.

Main Body:	
	By and between
	Recitals and definitions
	Transaction
	Conditions precedent
	Termination and withdrawal
	Consideration
	Closing and actions at closing
	Representations and warranties of the seller
	Representations and warranties of the purchaser
	Indemnity of the seller
	Confidentiality
	Applicable law
	Arbitration
	Miscellaneous
Annexes:	
	Power of attorney
	Description of the assets
	Data room
	Payment agreements
	NFP
	Reference date accounts
	Deed of transfer

TABLE 7.5   Example of Structure of a QPA Agreement

authorization to act on another's behalf, useful when the persons involved in the mechanics of the transaction do not have formal legal power of representation of the parties, that is, are not the managing director.

An example of a QPA structure is given in Table 7.5.

7.13   Sale and Purchase of Equipment Agreement

A SPA is the contract between two parties that rules the transfer of ownership of equipment or goods. One party, the seller, sells the equipment to the buyer (or purchaser), who purchases the equipment.

A section entitled *scope of purchase* provides the perimeter of the agreement, namely, the type of the equipment, or goods, with model number and reference to data-sheet in the annexes, with inclusions and exclusions of ancillary performances (like transport, delivery, labor, training. . .).

The *commercial terms* define the price, or consideration of the sale and the schedule of payments, which are normally requested as steps at signing of contract (10–40% of the total consideration), at delivery (10–40% of the total consideration), at final installation, and acceptance test (remaining of total consideration). Amounts that are not timely paid are normally requested to bear an interest rate which can be defined as monthly interest starting after a number of days from payment deadline, and the buyer will have to incur eventual costs and expenses incurred in collecting any amounts due. A payment security can also be called for, in the form of bank guarantees or letters of credit, normally from an international first-class banking institution when transactions are international and not local. A subsection defines the responsibility for taxes and duties that is borne by the purchaser for any sales, use, excise, value-added, services, consumption, and other tax on the provision of the equipment.

A section on *shipping and delivery* regulates the delivery terms of the equipment by defining the *Incoterms* rules (or *International Commercial Terms*) which are commercial terms published by the *ICC*. Examples of Incoterms that regulate shipping are defined for general as well as sea and inland seaways transports. Most used terms are the following:

- EXW (ex-works): the seller makes the goods available at its premises, whereby the buyer has to arrange for collection, and transport to the desired point of delivery

- FCA (free carrier): the seller transports the goods to the entity who is responsible for the subsequent transportation and delivery to the delivery point chosen by the buyer

- DPA (delivered at place): the buyer receives the goods directly at its own premises, but doesn't pay import duties and taxes

- FAS (free alongside ship): the seller delivers the goods alongside the buyer's vessel at the chosen port of shipment, loading, and subsequent operations being in the responsibility of the buyer

- FOB (free on board): the goods are under the responsibility of the seller only until they are loaded on the carrier, the buyer to carry the costs of transport, import duties, insurance, and unloading

- CIF (cost, insurance and freight): the goods are under the responsibility of the seller until they reach the delivery point and are unloaded from the carrier, together with the costs to be borne for insurance and transportation charges

Each of the Incoterms cause different cost allocation structures for seller and buyer. Part of the same section, usually added in the annexes, is the "packing list" (or "equipment list") that details all the equipment and its parts to be included in the delivery. Risk of loss shall pass to the buyer when the equipment is delivered to the delivery point and is not responsible for any damage, loss, or expense after delivery. Product, quantity, price, shipping destination, delivery terms, and delivery timing are listed as "commercial and other terms" under the annexes to the agreement, and the seller usually reserves the right to make partial deliveries and to submit invoices for such partial deliveries, while at the same time, reserving the right to allocate available equipment among its customers in its discretion.

A full section is devoted to the *warranties* where the seller warrants the *no liens* clause:

Seller warrants that all the equipment sold under this agreement will be free from any of any lien, claim, or encumbrance of any nature by any third person and that the seller will convey to company clear title to the equipment as provided for in this agreement.

The *limited warranty* clause warrants, on the part of the seller, that the equipment will conform to disclosed specifications or other specifications as the parties may agree in writing, provided that the equipment is subject only to the usage for which it is intended. Normally, the warranty starts at the time of delivery and continues for a period which normally can range between 1 and 2 years, or more if agreed in writing. In the event that any equipment fails to comply with the foregoing limited warranty, the buyer must, no later than expiration of the applicable warranty period, inform the seller who will replace or repair, at its sole option and expense, any defective equipment or parts that prove to be defective during the warranty period. The seller will usually ask that this is the sole and exclusive remedy available to the buyer, with the exception of the liquidated damages stated in other sections of the agreement. If the seller determines in its sole discretion that the defect is attributable to any cause other than poor workmanship or defective materials or to a cause not imputable to the seller, then the seller itself shall have no obligation with respect to repair or replacement of the defective equipment. The seller shall return the equipment to the buyer, with expenses borne by the buyer, and the warranty will become void. The limited warranty does not cover defects caused by normal deterioration and wear and tear, and does not apply if the equipment has been subject to modification, misuse, mishandling, misapplication, operation outside rated capacities, negligence, improper maintenance, or accident or if any adjustments or repairs have been performed by anyone other than the seller or an authorized service representative of the seller. At delivery, the equipment must be controlled and accepted by the buyer; if the equipment is found to be "nonconforming," the buyer has the right to ask for curing the missing or damaging parts, up to a new shipment with LDs to be paid by the seller.

In the section on *liability, indemnity, and insurance*, the seller agrees to indemnify and defend the buyer against third-party claims, damages, and losses that arise out of any material defects in the equipment or out of the negligent acts or omissions of seller provided that the seller is promptly notified. The seller usually requires to be given complete control of the defense and settlement of the claim, taking advantage of the buyer's cooperation who will also have the right to participate in the defense. Symmetrically, the buyer will have to hold the seller harmless against claims arising out of the negligent acts or omissions of the buyer with respect to its performance of the agreement, or out of the alteration or modification of the equipment by the buyer, with the buyer having the right to be given complete control of the defense and the seller cooperating with the buyer in the proceedings. During the term of the warranty, the seller shall obtain, maintain, and keep in full force and effect a general liability insurance including a product-completed operations endorsement with a reputable insurer of at least S&P's A rating and no less than an amount, normally in the range of 1–2 million EUR, in the aggregate for any and all claims of personal injury or property damage directly caused by equipment covered under such insurance policy. Lastly, the buyer shall, at the seller's cost, give any assistance that the seller shall reasonably require to implement a product recall, as a matter of urgency, of equipment found to be defective, inherently dangerous, or posing a similar risk to the public safety.

An important section is devoted to the regulation of the *IP*, namely, all patent rights, patent applications, rights to apply for patents, know-how, copyrights, copyright registrations, right to apply for copyrights, trade secrets, trademarks, trademark applications, rights to apply for trademarks, and any other confidential information or proprietary information owned by the seller and its affiliates. The first covenant is that the sale does not represent a transfer of the IP to the buyer (the *no conveyance* clause), but the buyer is granted a worldwide, royalty-free license to that IP required for the installation and use of the equipment (the *"limited license granted to company* clause). Seller must indemnify and hold harmless the buyer in case the equipment infringes upon any patent or other IPR, with the buyer promptly informing the seller in case any claim is brought against the buyer by a third party; in the event the buyer is enjoined from use of the equipment or part of it, the seller will have to bear all expenses to either procure for the buyer the right to continue using the equipment, replace the same with a noninfringing equivalent, or remove the infringing part or the equipment and refund the buyer of the purchase price and all other costs. Finally, reverse engineering, by any means, is prohibited by the seller unless with its prior consent.

The *limitation of liability* clause permits contracting parties to reduce or eliminate the potential for direct, consequential, special, incidental, and indirect damages should there be a breach of contract. In some cases, a cap on damages may be introduced.

The *termination* section contains clauses that are always subject to some negotiation before signing. The buyer can normally enter termination at any time and "not for cause." In case of such termination, the seller may request termination charges that can consider reimbursement of costs incurred by the seller to manufacture, market, and sell the equipment. The buyer can terminate the contract with no termination charges in case of equipment nonperformance and the seller not fulfilling its warranty obligations, delayed delivery after agreement's time limits, exceedance of the maximum liquidated damages, occurrence of an event which in the reasonable judgment of the buyer will materially and adversely affect the ability of seller to meet production, equipment availability, or delivery obligations including but not limited to, acts of war, acts of terrorism, the enactment by a governmental authority in any country or state in which the buyer or the seller do business of any laws, rules, decrees, regulations, or statutes.... Both parties can enter termination in case of failure to comply with any material obligation or provision of the agreement and if such failure is not cured within a specified time limit (usually in days), violation of applicable laws or regulations, trademarks, IPR, request by law, or one party submits fraudulent or intentionally erroneous reports or information to the other party, if the other party becomes insolvent or ceases to or ceases to be duly authorized to, conduct business.

The SPA is completed by the remaining usual contractual sections as already seen in the previous paragraphs. An example of an SPA structure is given in Table 7.6.

7.14   Distribution Agreement

Similar to the SPA, a *distribution agreement* (DA) is an agreement by which a manufacturer appoints a company the distribution rights of its products in a certain territory, with or without exclusivity rights. The main difference between the SPA and the DA is in the rights set forth in the *appointment* section, where the manufacturer can give the company the exclusivity to sell its products in a specific territory, asking the company not to sell competitor's products and not enter other territories not included in the DA list. Some obligations are entered into between the parties; for instance, the company must not represent itself as an agent of the manufacturer for any purpose but can only represent itself as a distributor of the manufacturer; the company will not have the right to pledge any manufacturer's credit, give any condition or warranty, make any representation or commit the manufacturer to any contracts, and not make any promises or guarantees

Main body:	
	Scope of purchase
	Commercial terms
	Shipping and delivery
	Warranties
	Liability, indemnity and insurance
	Liquidated damages
	Confidential information
	Intellectual property
	Limitation of liability
	Termination
	Obligations of the parties
	Independent contractors
	Force majeure
	No assignment
	Publicity
	Dispute resolution, arbitration
	Representations and warranties
	Severability
	Counterparts
	Modification, waiver
Annexes:	
	Technical specifications
	Packing list
	Commercial and other terms
	Draft of bank guarantee of letter of credit

TABLE 7.6 Example of Structure of a SPA Agreement

about the products beyond those contained in the promotional material supplied by the manufacturer.

In case of an exclusivity appointment, the manufacturer undertakes to supply the products only to the company for resale in the territory and to prevent, during all the exclusivity period, any direct or indirect sale of licenses, patents, or products to third parties, including customers. The company also undertakes not to produce (or make third persons or firms to produce) similar products to or reproduce (or make third persons or firms to reproduce) copies of the products, as long as doing so constitutes a violation of the IPR; in other words, the company could eventually manufacture similar products but without infringing any know-how and IPs of the manufacturer; sometimes, this clause is subject to a time limit of one or a few more years.

A *marketing and sales* is an addition to the SPA since the manufacturer needs to make sure that the company is active in the commercialization of its products after having received the exclusivity appointment, by using its best endeavors to promote the distribution and sale of the products in the territory, like advertising, promotion, and deploy a full-fledged marketing strategy to push the adoption of the technology in the market of reference. Part of the marketing and sales section is the sale of a minimum quantity which must be sold by the company every year. A regular reporting is requested by the manufacturer to monitor the advance of market efforts by the company, with the latter having also other additional obligations as to organize after-sales services or insure the products as they are stored in local warehouses.

References

1. Garner B. A. (Ed.). (2009). Black's Law Dictionary, 9th ed. West Group: St. Paul, MN.

CHAPTER 8

Financing

Summary. Especially in the case of new technology and new business models that are lacking track records of successful operations, securing financing for project deployment can be difficult if not impossible. A project, to be considered for financing and be placed in the condition to become reality, must be structured in a way to reduce uncertainty on technical feasibility and economic viability to a level deemed acceptable by the providers of capital. This chapter surveys the range of possible sources of capital and necessary guarantees on project operations to create the basic conditions for an energy storage project to be built and operated.

8.1 Introduction

The real-life realization of an energy project is fundamentally linked to the economic profitability of the investment. Especially in the case of new technology and new business models that are lacking track records of successful operations, securing financing for project deployment can be a prohibitive endeavor.

To overcome stalemate and obtain funding to proceed with construction, the project must result *bankable*, that is, have characteristics that make banks, or lenders in general, willing to provide funds in return to guarantees on the project technical and economic performance.

Long-term warranties on equipment performance are critical, with warranties that must be backed by storage technology manufacturer or EPC and O&M contractors. To prove the technology, pilot projects that demonstrate the goodness of the technology and its proper functioning are fundamental to overcome lack of knowledge and fear of uncertainty. Bank, or lenders, can call in technical advisors of reputable standard to analyze and certify the technology after assessment of its characteristics and faultless functioning.

To ensure project economic viability, the enforceability of the revenue streams and the creditworthiness of the off-taker are also critical. Banks and lenders must be assured that a solid contract will guarantee the stream of money that will make the energy storage business plan profitable and capable of servicing the debt contracted by the project owners with the lenders.

Therefore, a revenue contract, a good creditworthiness of the off-taker, and technical performance guarantees are essential to the deployment of energy storage projects. This chapter surveys the range of possible sources of capital and necessary guarantees on project operations to create the basic conditions for an energy storage project to be built and operated [1–3].

8.2   Sources of Revenues for Energy Projects

Energy projects can benefit from a number of potential sources of profits related directly to commercial and marketing activities, like the sale of electricity to final users, or the avoidance of costs stemming from taxation or to the daily operations of a commercial or industrial user.

8.2.1   Power Purchase Agreements

A *Power Purchase Agreement* (PPA) is the contract entered into by one party, the producer of energy (either a utility or an *independent power producer* (IPP), and a buyer of the energy. A PPA is normally a long-term contract that not only locks the buyer, but also the seller, to performing the sale of energy according to a set of requirements on the quality of service and the price paid for receiving the service. In project finance for example, a PPA is fundamental in determining the credit quality of the revenues generated by the project. Price definition, secure revenue streams, and length of contract are the key points for a good PPA contract.

One example of price setting for a PPA can be devised around the monetary benefits derived from the reduction of consumption of fossil fuels, basing the remuneration of the project on the estimation of the avoided costs. The mechanism of price definition is an effort to balance the different interests of the parties involved in the project: the seller, the financing institutions, and the buyer. Often, such PPA is validated, or even regulated, by local energy authorities.

The PPA is a bilateral contract between one seller, also called the IPP and the buyer, or *off-taker*. Buyers can not only be residential, governmental, industrial, or commercial customers, but can also be electricity traders that further distribute electricity to their own customers. Creditworthiness of the off-taker and the quality of the contract are prominent in guaranteeing the quality of the revenue streams and, therefore, securing the financing of the project from third parties. The PPA assumes a central importance especially when IPPs are looking for nonrecourse project financing (PF) to develop the project into a real production asset.

The sale of electricity can occur by means of a direct connection between the IPP and the buyer, or in a distributed connection where the IPP sells the electricity to multiple *points of delivery* (POD).

In a typical PPA, the *sale agreement* clause states that, according to the terms and conditions in the remainder of the PPA:

> . . . the buyer agrees to purchase the entire power and energy output of the project, net to the point of delivery, and the producer agrees to sell and deliver said power and energy solely to the buyer.

Purchase price clauses define how energy and power in many PPA contracts are billed by the IPP. The easiest way to quote the price of energy is to compute all costs for the generation of energy, add an industrial margin, and divide it by the amount of energy delivered to the buyer. Energy prices can, therefore, vary depending on variable cost conditions for the IPP. For instance, if generation is performed using fuels, in case of fuel price variations, the same can be reflected in the electricity prices charged to the buyer. Power is normally priced at a fixed amount which is multiplied by the power that is made available to the buyer. Energy price per unit of energy is often referred to as the *energy rate*, while power price on unit of power is called *power rate*, or *capacity rate*.

The section on *power and energy delivered by producer* sets forth the conditions on how the IPP bills energy and power to the buyer. For instance, bills are invoiced as monthly payments by applying the capacity rate to the metered power, and energy is billed at the energy rate applied to the total metered energy delivered by the IPP during the billing period of reference.

The *quality of service*, or *performance standards* section gives the minimum technical specifications that the service provided by the IPP must respect during the contract period. This includes, for instance, specifications on active and reactive power, grid frequency, downtimes, and black-start operations.

The section on *testing and commissioning* gives the specifications on how the project plant will be put in operations safely and respecting local regulations. The *commercial operation date* (COD) starts after tests are completed and the full acceptance of the project has been signed between the parties. Project operation period starts from COD until the end date as signed in the PPA.

The *term and termination* section provides the agreement on the duration of the contract, with further automatic extensions in the period of time with termination notice of 30 days prior to the expiration of the beginning of the additional term. The PPA is normally entered for a minimum period of time of 5 years and extend to 10–20, or even more, years. A PPA may be terminated earlier if the parties fail to meet their obligations set forth in the PPA. The IPP usually negotiates limitations on termination due to *curtailment* in the provision of its services to the buyer, whereby such curtailments in the power output from the project can become necessary due to power system interruptions, overloads in the distribution line, force majeure, or even changes in the electricity production costs. If energy and power provision is not possible due to problems at the IPP or buyer's site, damages are to be paid to the party that has caused the interruption of the service.

The *metering* section defines the type and technical characteristics of metering units, the formulas and constants to be used to convert metering readings to actual energy and power units, availability of suitable locations, sealing of the units, their remote datalogging, periodical testing and calibration. If metering accuracy decreases under 98%, energy and power can be metered by additional units (often installed by the customer) or on mutually agreed upon estimates.

The *change of ownership*, or *title* clause provides the terms for the transfer of electricity ownership between the IPP and the buyer; normally, the title of ownership of electricity is entirely transferred to the buyer after delivery at the POD. From POD downstream, electricity belongs solely to the buyer.

Depending on the type of buyer, if the IPP decides to sell the project, the buyer will receive a right of first refusal to purchase the project before it is offered to other third parties.

All O&M activities are agreed between the parties and detailed reporting on power generation records, frequency regulation, outages, management of protection devices, ordinary and extraordinary maintenance will need to be logged and made available for analysis. Finally, the usual set of clauses, common to most of other contracts, complete the PPA framework.

8.2.2 Energy Efficiency Certificates

A number of tradeable instruments have been proposed by regulators and implemented in the countries that have been more active in the field of climate change control.

The *green certificates* are tradeable titles that certify the production of energy from renewable sources on the part of the title holder. Typically, one certificate corresponds to the generation of 1 MWh only from renewable energy sources. Such certificates are exchanged in markets or by means of bilateral contracts, and their price can be negotiated according to demand and supply principles. The rationale underlying the trade of certificates lies in the fact that, if a firm that has not yet adopted a policy of reduction of its greenhouse gas emissions wants, or is requested, to purchase green certificates, such firm will eventually reduce its carbon footprint by creating a demand for green certificates therefore enhancing the economic profitability of other firms that has adopted a policy of environmental footprint reduction through the generation of clean energy. The demand would indeed increase the price of green certificates on the market and increase the market attractiveness, or the monetary value, of renewable energy technologies.

The number of certificates that can be obtained by producers of clean energy depends on the type of REN sources used, and the amount of energy produced with that source.

Due to the fact that the market for such titles is not necessarily very liquid and there can be large differences between the demand and the supply, a minimum price can be set by the regulatory authorities in order not to dilute the value of the title too much when supply is larger than demand. Regulators can also act by setting mandatory targets for the number of green certificates to be traded each year.

Typical participants to the green certificate market are REN generators, distribution companies or final consumers. A number of countries have started collaboration to create an international voluntary market for green certificates known as the *Renewable Electricity Certificate System.* (RECS).

The *white certificates*, or *energy efficiency titles* are titles that certify the saving of a quota of equivalent fossil energy, normally one tonne of oil, as a consequence of investments in energy saving initiatives. As per the green certificates, the white certificates are traded on a market, managed by national authorities, where demand and supply meet and prices are set, or by means of bilateral contracts. White certificates can be set at a fixed price over a period of time, normally 1 year, to facilitate the planning and execution of energy saving interventions.

Participants are energy-saving companies, distribution companies, or final users.

The *black certificates*, or *emission allowances*, are tradeable titles that grant the owner the permission to emit 1 tonne of CO_2 in the atmosphere over a specified period of time. They are valued at the the *marginal abatement cost* corresponding to the cost of sequestering 1 extra tonne of CO_2.

For the market of black certificates, typical participants are energy-intensive industrial users.

8.2.3   Feed in Tariffs

A *feed-in tariff* (FIT) scheme is a very popular and effective way to boost the introduction of REN plants in the energy system of a country. Many nations have adopted them with large success, often coming at some surprise to regulators when the FIT was set at very high levels. An FIT is the price paid per each kWh that has been produced with a renewable energy plant. Sometimes, FIT is indexed, but most often is a fixed contribution paid by the government over a period of several years (i.e., 20 years in case of photovoltaic plants).

An FIT can also be combined with the sale of electricity to the utility, who is obliged to buy at market conditions or at fixed prices set yearly by the regulatory body.

8.2.4 Carbon Tax

As an alternative to black certificates, the *carbon tax* is a tax levied on the carbon content of fossil fuel.

The value of this form of pollution tax, is computed accordingly to the *social cost of carbon* set equal to the marginal cost of emitting one extra tonne of CO_2. Such estimation poses some complexities due to the difficulties in understanding, from a scientific point of view, the residence time of CO_2 in the atmosphere, and its economic impact on society as an effect of climate change. Avoidance of payment of such taxes can be considered a source of revenues (avoided costs) for the introduction of energy efficiency or REN generation by industrial users.

8.2.5 Application-Specific Revenues

As seen in previous chapters, a myriad of applications are made possible by the use of ESS. All of them can be remunerated depending on local market regulations, and many of them will be valued in monetary terms in the near future with the increasing penetration of energy storage technologies.

The capability of ESS to regulate voltage and frequency of the grid to which they are connected is, in certain countries, recognized by the utilities as a fundamental service and remunerated under capacity payment contracts.

The use of an ESS as backup power can be one of the major sources of revenues in a project. If short, or even long, interruptions in the energy supply from the grid are frequent, and each one of the interruptions causes serious disruptions to an end user's operations, the avoidance of such costs is a fundamental service that remunerate an ESS and increase its profitability. The improvement of the quality of service of the energy supply can be, in some instances, a fundamental need of a customer that can be met only by energy storage.

When utilities bill the price of electricity, not only the cost of generation is charged but also a series of ancillary costs are transferred to the end customer. For instance, costs for metering services, taxes, and other forms of government payments (i.e., *excise duties*), and costs related to the transmission and distribution of electricity over the grid. Depending on the power draw that each customer requires from the grid, a sort of rental fee is charged to the end user: the higher the power drawn by the load, the higher the charges (normally as a fixed price per kW). If energy storage is capable of capping the power by providing energy to the load when a certain power threshold is surpassed, the end user will reduce its amount of fixed charges per kW. Depending on the type of load and ESU energy content, the *power draw reduction* can be one additional stream of revenues for an ESS project.

Time-of-use, as discussed in a previous chapter, enables the flow of revenues by the arbitrage on electricity prices that differ during different hours of the day.

Energy from REN, and managed by storage systems, can benefit from a priority on the energy injection in the grid with respect to electricity converted from other forms of energy sources. The merit order energy dispatching is the concept that energy produced at lowest or zero marginal costs has priority of dispatch against energy from fossil fuels, that have higher or nonzero marginal costs due to the consumption of fossil fuels, at their cost, in the generation plants. This way, energy power plants that has dispatch priority can survive during periods of low energy prices against power plants that operate at much higher marginal costs. Although not a revenue stream per se, this is viewed favorably by investors that can count on the plant being always allowed to extract value from grid connection, since it can always injection energy in the grid.

Most of the applications that stem from the deployment of ESS have an economic value, but its recognition is heavily dependent upon local regulations and interplay between centralized and decentralized players in the energy market, namely, between monopolists or oligopolists and the newcomers. Nonetheless, such value is manifest and obvious, and should be rewarded according to its utility for the final consumer.

8.2.6   Other Agreements

Agreements that are common in the energy industry, mainly in the provision of gas supplies, can be considered for the construction of alternative agreements for electric energy contracts based on energy storage technology.

Examples of such agreements are:

- the *supply-or-pay* agreement, entered when one party, the supplier, is made accountable to deliver the object of the supply or else pay for the assured quantity.

- the *take-or-pay* agreement, a contract between a supplier and an off-taker that either gets the product from the supplier or pays a penalty even if the object of the supply is not taken, usually with a cap.

- a *take-and-pay* agreement, the direct supply for the payment of the amount where the off-taker takes all the contractual quantity of energy and pays accordingly.

- a *throughput agreement*, the provision of a specified amount of energy per period to a distribution line or pipeline.

8.3   Equity

Equity is the capital that is injected in the firm by the owners or its stakeholders. It represents capital that is fully at risk: if the firm defaults, the money is lost. Whoever provides capital to someone else, wants something back: thus equity comes at a cost k_e, the cost of equity, which is a percentage with values typically over 10% per annum (see Chap. 6).

When the owners inject capital in the firm, they get a title of ownership that depends on the percentage of their injected capital over the total equity of the firm. As a title of ownership, shares are attributed to the equity participants according to the percentage of their equity injection. Shares can be:

- *Ordinary shares* are the most senior form of ownership. These shares have direct impact on the firm's ownership structure and management, since they grant to the owner the right to participate in top-level decisions by carrying voting rights without limitations. Their value depends directly on the revenues generated by the firm's operations, and materialize as dividend payments or capital gain when the shares are sold.

- *Preferred shares* maintain ordinary share basic rights, with limitations set on the decision-making process, but with additional rights granted to minority investors to protect their stake in the firm. The preferred shares can anyway grant their owners with certain voting rights up to a limit, typically around 25%, allowing the holders to veto some resolutions in specific occasions as defined on the shareholder agreement. Rights conferred to preferred shares can be:

- *participating rights*, offering the right to receive dividends if the firm is profitable over the period

- *cumulative rights*, for which if the dividend is not paid, it accumulates for future payment, and paid before all other shares

- *preferential rights* which grant the the right of receiving dividends over ordinary shares on firm liquidation, and also carry additional voting rights with respect to some kind of decisions to be taken by the management

- *convertibility rights* shares can be converted into ordinary shares, normally at trade sale or flotation

• *Redeemable preference shares* do not carry the right of conversion into ordinary shares, as they are more intended to increase financing by acting as a funding instrument alongside ordinary shares. They can be vested with the right to receive a coupon or dividend or be redeemed at a premium over the initial price.

Stock options is the provision of stocks to the management team or even personnel without direct responsibilities of the firm's operations. When stocks are granted to employees, the investors accept to reward management for the firm's performances by allocating an additional number of shares, or by using shares that had been reserved and set aside specifically for this purpose.

Mezzanine capital is another form of equity injection. Normally, mezzanine financing can be structured either as unsecured and subordinated debt, or as preferred stock. Mezzanine capital is often a more expensive financing source for a company than secured debt or senior debt, its higher cost of capital being due to its nature of unsecured and subordinated (also termed ad "junior") obligation in the firm's capital structure. Due to its nature, if the firm defaults, it is repaid only after the payment of other more senior obligations, but only if there is enough money left to service the mezzanine. The use of this financing instrument increases the firm's leverage and, therefore, increases the financial risk of the firm.

Apart from the capital gain that materializes when shares are sold at trade sale or flotation, equity is remunerated by the *dividend and distributions policy* as defined in the shareholder agreement. Typically, shares are entitled to be paid a dividend if so decided by the board of the firm. Dividends are paid from the "cash available to distribution" saved into the *distributable reserves*.

8.4   Private Equity, Venture Capital, and Financing Rounds

Private equity (PE) funds are institutional investors that buy shares of a target company in order to participate in the life of the firm, increase the value of their shares, and exit at a premium over, normally, the medium term. *Venture capital* (VC) funds acquire shares of firms with high growth potentials in sectors that promise to develop successfully in the following years. Such firms are anyway still in the early stages of their life cycles, but that are still in their start-up phase, and as such fail to attract capital from banks or exchange markets. Normally, PE and VC define their exit strategies already when evaluating entering the equity of a firm. Normally, such strategies can be the *trade sale* of the firm, management buy-ins and buy-outs, or an *Initial purchase offer* (IPO), The typical period of time for them to remain in the capital of a firm being in the range of 4–5 years.

The financing of a new firm can be schematized, most of the times, as several, subsequent rounds that occur after milestones in the life of the firm are reached. The initial equity injection typically is referred to as *seed financing*, and represent the capital provided to perform the initial development from the idea, to the first prototype, to initial patent filing, first market analysis, and business planning. The goal is to understand, and prove if there is a business case for the idea. At this stage, normally only the founder's money is invested since risk is too high for external funding to be raised. At this stage, another possibility for external funding can be represented by *angel investors*, who acts as an alternative to more established VC firms. Angel investors are normally experts in the field, know the business space well and are confident that there is a business case for the founder's idea. The angel investor can follow the operations of the start-up as an external advisor, or decide to be more active in the management of the firm. A very good feeling between the founder and the angel investors is often the main decision point made by the founder whether to take on board an angel investor, aside from the liquidity capacity that the angel investor is capable and willing to put into the firm.

When the business case is demonstrated and more confidence can be granted to the potentiality of the firm, *start-up financing*, or *series A financing*, can be raised from its first outside investors, or venture capitalists. At this stage, the firm is incorporated and can see some initial revenues starting to flow in. The money is used to launch the business idea in the market, perform branding and deploy its marketing plan, finance its working capital, research and development, first hiring of people, rent of building, and deployment of the first manufacturing capabilities. Series A financing normally comes in the form of convertible shares that become common stock at IPOs, management buy-ins or buy-outs. Series A rounds are a critical stage in the funding of new companies, and typically lead to the dilution of 10–30% of the company's equity. Series A valuation, namely, the money raised at this stage of financing, is made at an amount needed to capitalize the company for 6 months to 2 years of operations.

Series B financing comes after the successful launch that has proven the viability of the business case. The money is used to further penetrate into the market, establish a position into the competitive arena, enlarge and consolidate operations, fund more working capital, and hire more experienced management and professionals. Several subsequent financial rounds, that can be called *series C, series D* financing can be flown in according to milestones that are reached during the following phases until monthly cash flows become stable, break-even is reached, and the firm can start considering the use of bridge financing, mezzanine debt, or bank loans to establish its operations and reach the maturity phase. No VC or PE are needed at this stage.

At this stage, VC and PE investors can start considering pursuing the exit from the firm.

The founders must pay attention during each round, when the company is valued and new investor's money brought in. As a consequence, too many rounds can overly dilute the owners' stakes in the venture. It is also important for the founder not to enter a IPO too early or when the market conditions are not favorable, to avoid losing value to other investors' benefit.

8.5   Debt

Debt is the amount of money borrowed by one party, the borrower, from one or more lenders. For his service of lending money, the lenders require the borrower to pay an interest i, which represents the cost of debt k_d (see Chap. 6).

Debt differs from equity as it is normally a fixed claim against a residual claim of the equity, has high priority over cash flows, is tax deductible, and has a finite life (a fixed *maturity*) as opposed to the infinite life of equity.

An indebted firm can have advantages over firms that are all financed with owners' equity only: apart from the tax benefits, debt imposes discipline on the part of the management team, who is forced to service the debt and, therefore, better manage the cash flows of the firm. As a trade-off, debt reduces the flexibility of the firm to find further financing as too much debt would not be tolerable and would prevent further lenders to provide capital when the debt over equity ratio is considered to be too high.

The right amount and type of debt is a thus a key issue for an entrepreneur, and can make the difference, many times, between a solid company and a fledgling organization that has difficulties in managing its mix of debt and equity. Using debt alongside, equity can make the entrepreneur better off in case of net profits. The *financial leverage*, or *gearing*, represented by the debt to equity ratio, not only increases profits when the returns from the firm's operations are larger than the cost of borrowing, but also losses are increased if the opposite is true. Having a too high gearing can pose threats of bankruptcy in firms that are too leveraged and cannot service the bank debt in case of a downturn.

8.5.1   Loans

Debt usually comes in the form of a loan, which can be unsubordinated (or *senior debt*), meaning that the repayment of such debt will occur prior to any other payment from the borrower, or subordinated (or *junior debt*), meaning that such debt ranks after other debts according to the repayment schedule of all other debts held by the firm.

Loans can also be categorized according to their scope and time span; for instance, a *bridge financing* loan is used to cover operating costs like the ones occurring during project construction, a *stand-by financing* loan covers cost overruns, a *working capital facility* is a revolving loan that is intended to cover working capital during a period of time[1] *Syndicated loans* are a form of debt provided by a group of financial institutions that aim at sharing the risk when the debt amount is large and is financing large projects.

The loan amount (the *facility*) is repaid over the loan duration period, or *tenor*, according to an amortization schedule: the principal and the interest portions of the loan are repaid over the life of the loan (in banking terms, is "amortized") according to different methodologies.

The life of the loan is monitored using a set of ratios. The *debt service cover ratio* (DSCR) is the ratio of the *cash flow available for debt service* (CFADS) over debt service payments (interest and capital repayments) calculated at each interest payment date as follows:

$$\text{DSCR} = \frac{\text{CFADS}}{\text{Debt Service}} \tag{8.1}$$

---

[1] A *revolving loan*, or *evergreen loan*, allows the loan amount to be drawn down, repaid, and drawn down again, in any amount and timing desired, until expiration, of the agreement.

CFADS are the cash flows that can be computed as the free cash flow to the firm as in Sec. 6.6.

DSCR is used to check if the financial resources generated by the project are sufficient to maintain coverage of the debt service. It can be computed as an annual DSCR (ADSCR) and used, for the purpose set forth in the loan agreements, as ex ante (historical) or ex post (forward-looking) based on forecast assumptions and periodic reviews. Another DSCR ratio can also be computed as the average DSCR over the years of debt service.

The *loan life cover ratio* (LLCR) is the sum of the net present value of the *operating cash flows* (OCF) from the moment of valuation to maturity, also taking into consideration the reserves accrued to service the debt (the DSRA, as explained later in this section) over the net present value of the debt outstanding at time of valuation; OCF can be computed similarly to CFADS as the free cash flow to the firm as in Sec. 6.6:

$$\text{LLCR} = \frac{\sum_s^n \frac{\text{OCF}_t}{(1+i)^t} + \text{DSRA}}{O_s} \tag{8.2}$$

where s is the year at LLCR valuation, n is the maturity year, $t = s, \dots, n$, OCT_t is the operating cash flows at year t, and O_t is the net present value of the outstanding debt at time of valuation. The LLCR, if greater than 1, represents a valuable piece of information for the parties, since it means that the available cash flows generated over the remaining loan period would be able to more than cover the outstanding debt until maturity if the borrower were to close the loan at time s.

Both DSCR and LLCR can be used as triggers during the life of the loan. For instance, if the DSCR or LLCR become less than a specified threshold (i.e., 1.2, the *trigger ratio*) during the life of the loan, the loan agreement will set forth some provision in order to enforce the return of such ratios over such specified thresholds.

One loan-repayment method that is commonly employed for repayment is the *constant amortization schedule*, in which the principal repayment is maintained constant over the duration of the loan, while the interest payment decreases according to the decreasing portion of principal that is repaid over the duration of the loan.

If:

- T is the number of periods over which the loan is to be repaid with $t = 1, \dots, T$ (T can typically be monthly, semiannual, and yearly)
- L_t the beginning balance of each period of the residual loan amount
- P is the principal repayment
- I_t is the interest payment over period t with i the interest rate

the following amortization schedule can be computed. The principal repayment for each period is:

$$P = \frac{L_1}{T} \tag{8.3}$$

The beginning debt balance per period is given by:

$$L_t = L_{t-1} - P, \quad t > 1 \tag{8.4}$$

Period	Beginning balance	Payment	Constant Principal	Interest	Ending balance
1	100,000.00	−10,000.00	−5000.00	−5000.00	95,000.00
2	95,000.00	−9750.00	−5000.00	−4750.00	90,000.00
3	90,000.00	−9500.00	−5000.00	−4500.00	85,000.00
4	85,000.00	−9250.00	−5000.00	−4250.00	80,000.00
5	80,000.00	−9000.00	−5000.00	−4000.00	75,000.00
6	75,000.00	−8750.00	−5000.00	−3750.00	70,000.00
7	70,000.00	−8500.00	−5000.00	−3500.00	65,000.00
8	65,000.00	−8250.00	−5000.00	−3250.00	60,000.00
9	60,000.00	−8000.00	−5000.00	−3000.00	55,000.00
10	55,000.00	−7750.00	−5000.00	−2750.00	50,000.00
11	50,000.00	−7500.00	−5000.00	−2500.00	45,000.00
12	45,000.00	−7250.00	−5000.00	−2250.00	40,000.00
13	40,000.00	−7000.00	−5000.00	−2000.00	35,000.00
14	35,000.00	−6750.00	−5000.00	−1750.00	30,000.00
15	30,000.00	−6500.00	−5000.00	−1500.00	25,000.00
16	25,000.00	−6250.00	−5000.00	−1250.00	20,000.00
17	20,000.00	−6000.00	−5000.00	−1000.00	15,000.00
18	15,000.00	−5750.00	−5000.00	−750.00	10,000.00
19	10,000.00	−5500.00	−5000.00	−500.00	5000.00
20	5000.00	−5250.00	−5000.00	−250.00	0.00

FIGURE 8.1 Example of a constant amortization schedule of a 20-year loan at 5% interest, 100 k amount.

The interest repayment at period t is given by:

$$I_t = i \times L_t \tag{8.5}$$

The period payment R_t, is:

$$R_t = P + I_t \tag{8.6}$$

An example of a constant amortization schedule is provided in Fig. 8.1.

The *annuity (or mortgage) amortization schedule*[2] method has periodic installments R that are constant over the period and given by:

$$R = L_1 \times \frac{i}{1 - (1+i)^{-n}} \tag{8.7}$$

and principal payments by:

$$P_t = R - i \times L_1 + i \times \sum_{h=1}^{t-1} P_h \tag{8.8}$$

---

[2] This schedule is also referred to as "french amortization" method.

Period	Beginning balance	Annuity or Mortgage Payment	Principal	Interest	Ending balance
1	100,000.00	–8024.26	–3024.26	–5000.00	96,975.74
2	96,975.74	–8024.26	–3175.47	–4848.79	93,800.27
3	93,800.27	–8024.26	–3334.25	–4690.01	90,466.02
4	90,466.02	–8024.26	–3500.96	–4523.30	86,965.07
5	86,965.07	–8024.26	–3676.01	–4348.25	83,289.06
6	83,289.06	–8024.26	–3859.81	–4164.45	79,429.26
7	79,429.26	–8024.26	–4052.80	–3971.46	75,376.46
8	75,376.46	–8024.26	–4255.44	–3768.82	71,121.02
9	71,121.02	–8024.26	–4468.21	–3556.05	66,652.82
10	66,652.82	–8024.26	–4691.62	–3332.64	61,961.20
11	61,961.20	–8024.26	–4926.20	–3098.06	57,035.00
12	57,035.00	–8024.26	–5172.51	–2851.75	51,862.49
13	51,862.49	–8024.26	–5431.13	–2593.12	46,431.36
14	46,431.36	–8024.26	–5702.69	–2321.57	40,728.67
15	40,728.67	–8024.26	–5987.83	–2036.43	34,740.84
16	34,740.84	–8024.26	–6287.22	–1737.04	28,453.62
17	28,453.62	–8024.26	–6601.58	–1422.68	21,852.05
18	21,852.05	–8024.26	–6931.66	–1092.60	14,920.39
19	14,920.39	–8024.26	–7278.24	–746.02	7642.15
20	7642.15	–8024.26	–7642.15	–382.11	0.00

FIGURE 8.2 Example of a annuity/mortgage amortization schedule of a 20-year loan at 5% interest, 100 k amount.

Interests are given by:

$$I_t = R - P_t \tag{8.9}$$

An example of an annuity/mortgage amortization schedule is provided in Fig. 8.2.

A number of other amortizations schedules can be devised for loan repayments. For instance, a *target DSCR amortization schedule* is devised as to maintain the DSCR constant over the loan period; a *sculpted amortization schedule* is planned in advance and laid out as a table of repayments according to negotiations between the parties without the need for a formula to compute each installment.

A term sheet for a bank financing introduces the borrower and the lenders together with the purpose of the lending and the total cost of the investment, the definition of the facility, its amount, the tenor, the maturity, the eventual *grace period*, during which if the first payment is not undertaken, no late fees will apply, the type of amortization schedule of the loan, and the drawdown mechanics. The first draw-down is usually bound to the respect of a set of condition precedents, like having in place all project completion certificates, certification that all project requirements have been met, deed of assignments for the cash-flow streams, shareholder loans covering, for instance, VAT and grace period interests.

A clause defines the interest rate that is usually the sum of the floating interest rate (i.e., Euribor or Libor), the spread depending on market conditions, and the interest rate swap to cover the floating interest risk.

The cost and expenses section of the term sheet defines the bank fees that usually consist in an upfront fee in the range of 40–120 basis points calculated on the loan amount, payable for instance in two tranches (i.e., 20% at signing of the application and 80% at contract signing).

A set of *control accounts* receives the operating cash flows of the project and enables the lenders to verify and control the use of such proceeds. The lenders can, therefore, exercise control over the use of such accounts by monitoring the origins and uses of cash and retain a preemptive authorization right on all proposed expenses requested by the borrower for the asset management.

The *proceeds reserve account* host all project revenues that can be withdrawn for operating purposes or dividend payments, after previous approval from the lenders.

A *compensation reserve account* is used to receive compensation payments from damages, insurances, and a *maintenance reserve account* is employed to create a reserve used to cover ordinary or corrective maintenance of the project assess.

A *working capital reserve account* is used to create a buffer to cover running costs, and a *disbursement account* receives the drawdowns and eventual equity subscriptions. Withdrawals would be permitted to fund construction costs.

The *debt service reserve account* (DSRA) is a fund that guarantees sufficient money to service debt when the cash flows from the project are lower than budgeted. The DSRA can be equal to 6 months of debt service, but its balance is given back to the borrower at the date of the last interest payment. Normally, the DSRA remains under the control of the borrower, but the bank where the account is kept is typically required to have a very high rating (i.e. S&Ps A1).

The *cash trap* reserve account is used when the DSCR and LLCR ratios have been triggered. In that case, all revenues from the project net of operating expenses, including taxes, and debt service payments, are transferred to such account until the triggers, and all other control ratios, have been reset to the levels set forth in the loan agreement. At that moment, the funds are released from the cash trap account in accordance with the waterfall policy of the loan agreement.

The *waterfall payment structure* gives the priority of payments that are credited during the life of the loan; the sequence is normally set forth as the following:

1. agency fees
2. operating expenses
3. operating reserve accounts
4. taxes
5. debt service costs
6. DSRA
7. cash trap
8. distributions, dividends

Distributions are typically permitted if all required payments are made to the DSRA account, the DSCR is maintained over the minimum level, and there are no events of default.

A set of guarantees are requested by the lenders to the company as already seen in Chap. 7. Guarantees can be, for example: a deed of assignment from the company to the lenders of the receivables arising from the operations of the project; a mortgage on the land or buildings; special pledges on movable assets depending on local regulations; assignment of the the rights arising from the insurance policies; the respect of the limits on the average DSCR (normally not lower than 1.2) or amounts accrued in the DSRA, and D/E ratios under the 90/10 threshold; the distribution of dividend to be permitted in compliance with all covenants and obligations of the loan; commitments from the shareholders to provide coverage in case of revenues are lower than budgeted.

Normally, a *security package* is requested by lenders as a defense mechanism that is enforced to prevent other creditors to claim the project assets, and to enhance control so that the lenders can acquire direct control over the project to continue operations in case the main sponsor defaults in an attempt to receive the loan back. Thus, the security package can entail pledges over specific tangible assets or can be contractual agreements forcing the borrowers over specific performances or to avoid creating encumbrances over the project assets in favor of other parties. Other protections can be the enforcement of mortgages on fixed assets (i.e., land and building) or charges on movable assets (i.e., equipment), the request for project insurances, the negotiation of supply or off-take contracts, the assignment of the rights of permits or contracts as well as ownership rights over the project company (like the acquisition of shares with dividend rights).

Alongside with the security package, borrowers also provide the lenders with *covenants*, namely, undertakings to ensure that debt service does not deteriorate during the course of the loan tenor. Some covenants entail, for instance, that restrictions are placed on capital spending, new investments, main financial rations, like debt coverage ratios, on cash flow, on gearing, on raising additional debt finance, either with an equal or more senior claims, or distributing too much of the company's profits, the provision of periodical financial reporting. In case of a covenant breach, the lender can consider the loan agreement terminated and request that the debt be repaid immediately.

Other reasons for default can be the failure to pay the loan installments or to fulfill other obligations in the loan agreement, having declared untrue representations and warranties, cross-defaults on any other agreements, even in case of material adverse changes.

A hedging on interest rate from floating to fixed is normally required, to be entered a number of days after signing to provide coverage to interest rate risk exposure, on terms that must be acceptable by the lenders (the *hedging policy*).

At the end of the term sheet are the usual clauses found in contracts regarding nomination of advisors (technical, legal, and insurance, tax), exclusivity, costs, taxes to be paid by the borrower, governing law, and jurisdiction.

A structure of a typical bank financing term sheet is outlined in Table 8.1.

8.5.2   Bonds and Minibonds

A *bond* is a debt security[3] that corporations, banks, and government can issue in order to obtain fresh capital in the form of debt. The organization that issues the bond, receives from the bond holder money (the *principal*); in exchange, the issuer pays the bond holders an interest (the *coupon*) and repays the principal at the maturity date. Interest is usually due at semiannual or annual intervals.

---

[3] A *security* is a financial instrument that can be traded in institutional exchange markets.

Borrower
Purpose
Facility
Facility amount
Final maturity
Repayment
Drawdowns
Voluntary prepayments
Interest rate
Cost and expenses
Control accounts
Waterfall
Guarantees
Security package
Covenants
Default
Hedging policy
Exclusivity
Costs and taxes
Governing law, jurisdiction

TABLE 8.1   Structure of a Bank Loan Term Sheet

Normally, the coupon is is fixed during all the tenor of the bond, but it can also be made variable according to market indexes like the Euribor or Libor. The bond holder receives an interest rate and receives a reimbursement or premium at the maturity date, which provides the holder with a monetary gain, or *yield*.

When the issuer is not a financial institution but a firm, the bond is called *corporate bond*, since normally only large companies, or corporations, can be permitted the emission of bonds. For smaller firms, it is possible to proceed to the issuance of *minibonds*. Just like normal bonds, the minibond has an interest rate and a date of maturity. The difference between minibonds and classic corporate bonds is the fact that they can be issued by small to medium firms without necessarily resorting to banking or financial institution for intermediation. Normally, firms that want to issue minibonds must have a turnover in the range of 2–5 million (i.e., euros), margins in the range of 10%, certified financial statements for the last 1 or 2 years, but do not need to have a rating or to be listed in the stock exchange. Only institutional or qualified investors can subscribe to the minibonds, while other investors can only purchase such bonds on the secondary markets that banks organize for their trading. The minibonds are interesting to small firms also because issuance costs can be very low and there is no need to be advised by banks or financial institutions. Typically though, banks offer advisory services at very low prices, and can offer minibonds of different firms bundled together as a new category of debt instruments. Yields for the investors are normally higher than corporate bonds, due to inherent higher riskiness of the asset due to their low liquidity and high volatility.

An example of a term sheet for a bond or minibond transaction has a structure similar to senior debt issuance term sheets, which usually start by declaring the purpose of the loan and the definition of who the lender and the borrower are. Other terms concern the total money on loan, which kind of payment is applying (i.e., annual payment methodology), the set of related clauses [i.e., operational expenses linked to inflation according to the *consumer price index* (CPI)], the interest rate at which the loan is provided, the definition of the type of facility (usually of the *amortizing term loan*), the principal repayment schedule, the number of payments per year, and the loan maturity date. Also, the *day count convention* that determines how the interests are accrued over time. Drawdown timing of the facility is also stated, being it usually at closing. Repayments or use of proceeds deriving from the subscription of the bond and from the project operations are requested to reimburse in full the loans.

The *make-whole* provision, or *early redemption*, agrees that the issuer can redeem the bond in whole, but not in part, on any date of payment. Usually, penalties are applied to the issuer according to the year in which the early redemption is performed, decreasing yearly from, that is, 8–10% of the outstanding principal amount and reaching zero after 7–10 years from bond issuance.

The bond is defined in the term sheet to make sure that it ranks in right of payment as a "direct, senior, secured, and unsubordinated obligation of the issuer," alongside the existing and future senior debt of the company.

The set of documents that need to be negotiated and signed for the minibond to be issued is also listed in the term sheet, such documents being typically the direct agreement, the agency agreements, the borrower loan agreements, with other documents as requested by the bank.

In case of syndicated loans, the arranger bank or financial institution is provided in the term-sheet together with its arranging fees (i.e., 2% of total loan amount). Agency fees are also listed and can amount to, for example, 0.2% of the loan amount and being CPI-linked.

The usual set of control reserve accounts are put in place as described in Sec. 8.5.1. The DSRA, operating expenses reserve account, the working capital reserve account, and the cash trap account, together with the provisions on DSCR and LLCR triggers.

As per the loan-repayment structure, the waterfall method also apply for the bond and minibond loans, with the same or similar provisions on the distribution to equity holders.

The conditions precedent for the minibond to be granted are based on the results of the technical, legal, insurance, and tax due diligence reports, that verify the evidence of all permits, legal rights, contractual obligations, revenues origin enforceability, technical quality of the operations, and soundness of the financial model. Also, the full set of representations and warranties must be in place as a condition precedent for bond issuance.

The security package is intended to provide guarantees over the repayment and reimbursement of the bond. For example, the security package can request a pledge on the quota of the firm, a pledge on the firm's credit accounts, a first lien mortgage over the firm's property, special privileges depending on local regulations, and the assignment of credits from PPA or FIT contracts.

Covenants are contractual obligations that pose limits on the borrower in order to protect the lender from potential damages brought about by borrower's misbehavior. Examples of covenants are the contractual obligations ensuring that all the revenues shall credited to the accounts specified in the term sheet, all insurances be maintained valid during the full period of the agreement, keep in force of the validity of all mandatory licenses to operate the firm, maintenance of all representation and warranties, respect of all contractual obligations in contracts like the EPC, O&M, PPA... One important covenant is that a limitation is placed on further indebtedness on the part of the borrower, details of which are negotiated between the parties. Covenants are also placed on DSCR and LLCR, for instance, by setting their maximum value at 1.1–1.2, and leading to contract termination if such ratios are breached and not remedied. Covenants are tested (*covenant testing*) normally two to three times per year or at loan repayments. *Negative covenants* are covenants based on "shall not" behaviors, like for examples not paying dividends unless in compliance with the waterfall structure, not redeeming capital stock and making certain investments other than the authorized investments for the benefit of the operations of the project, changing the nature of the business, carrying out the transfer and sale of certain assets, or performing extraordinary corporate transactions, mergers or consolidations with other firms.

Failure to respect the contractual covenants can cause the loan to be withdrawn or its conditions to be changed to less favorable ones.

As default events, the term sheet considers the failure to pay the sums at the moment when they are due from the borrower to the lender, the breaches of DSCR, LLCR, a *cross-default* (meaning failures in other clauses that amount to a total cap when summed together) greater than a negotiated amount of money), insolvency, bankruptcy, and business cessation of the company, even if caused by a major damage to the operations that will likely cause the firm to go bankrupt as a consequence of such event.

During the life of the loan, the firm will be required to provide the investors with a set of documents like periodic reports on finance and operations, annual financial statements, and the analysis of variance between budgeted and actual results.

At the end of the term sheet are the usual clauses regarding nomination of advisors (technical, legal, insurance, and tax), the definition of which market the bond will be listed, exclusivity, costs (this time to be borne by the borrower, even for costs incurred by other parties in the context of the minibond issuance), taxes, governing law, and jurisdiction. A typical structure for a bond or minibond term sheet is provided in Table 8.2.

8.6   Lease Financing

Lease financing, or *leasing*, is an alternative way to bank loans since leasing can be more convenient, quicker to obtain and easier to structure. The entity offering the lease financing is called the *lessor*, while the entity receiving the financing is called the *lessee*.

Leasing comes in two forms: capital leasing and operating leasing.

Capital leasing, or *financial leasing*, is basically a loan from a financial institution. The lessee normally chooses and finds the asset that he needs, enters into negotiation with the lessor who ultimately buys the asset, provides the asset to the lessee for an agreed-upon payment schedule, and maintains the ownership of such asset until the last payment is made. At this point, the asset ownership is transferred to the lessee for a nominal amount or a bullet payment; each payment made by the lessee to the lessor is, therefore, a payment for an ownership portion of the asset that, over time, accrues until the asset becomes property of the lessee. A capital lease is structured as a loan, with an

Purpose
Lender, borrower definition
Loan:
Loan amount
Interest rate
Facility type
Principal repayment schedule
Loan maturity date
Payment dates
Interest rate swap
Day count
Drawdown availability
Repayments, use of proceeds
Make-whole (early repayment) provision amount
Bond ranking
Requested documentation
Arranger and arranger fee
Agency fee
Reserves accounts:
Debt service reserve account (DSRA)
Opex reserve account
Working capital reserve account
Cash trap reserve account
DSCR, LLCR triggers
Debt service coverage ratio (DSCR)
Debt service
Distributions
Payment waterfall
Conditions precedent
Security
Covenants:
Testing
DSCR, LLCR covenant
Property covenants and other covenants
Events of default
Required information
Advisory:
Technical advisor
Legal advisor
Insurance advisor
Tax advisor
Listing
Exclusivity
Costs and taxes
Governing law, jurisdiction

TABLE 8.2 Typical Structure of a Bond or Minibond Term Sheet

equity injection of typically 25–30% upfront payment of the nominal amount and a debt-repayment schedule as a normal loan.

The structure of a capital lease payments can entail the following series of disbursements:

- agency fees
- activation fees
- insurance premium
- deposit (equity injection)
- number of prepayments (equity injection)
- monthly payments from initial month to end of period
- bullet payment at end of period

The upfront deposit and prepayments represent the equity injection of the capital lease. Depending on local regulations on capital leasing, the item that is capital-leased must be treated as an asset on the balance sheet, and payments must be amortized as capital repayments and interest expenses.

Operating leasing is a rental contract for an asset, where the asset is owned by the rental company and maintained often with the help of a specialized third party, normally, a service or an asset-management company. The rented equipment is not recorded as asset on the lessee's company balance sheet and payments are made by the lessee with the advantage of being treated as input in the profit and loss statement, not as liabilities in the balance sheet. The rented asset is owned by the lessor, but can be transferred at the end of the leasing period to the lessee for an amount.

The structure of operating lease payments can entail the followings series of disbursements:

- agency fees
- activation fees
- insurance premium
- monthly payments from initial month to end of period
- payment at end of period

An operating lease can be structured with monthly installments (or rents), normally, for a number of years from 5 to 10, no deposits, interest rates depending on the project and macro-economical conditions, activation costs, insurance premium around 1.5–2% of loan amount, repurchase costs around 1% of loan amount. The item to be leased often must be scrutinized and accepted by the leasing company in a process called *accreditation*.

The agency and activation fees, also called *origination fees*, can be charged to the lessee, if not from the lessor, from a mediator, or broker that has found the leasing company and introduced it to the lessee.

Care must be exercised to fully acknowledge leasing costs since they can be somehow hidden in the term sheets and the contracts that regulate them; not fully understanding how a leasing is construed can bring about additional costs than can be avoided by careful negotiation or by cherry-picking between leases offered by different financial institutions. The payment structure can indeed make the leasing to become somehow difficult to be analyzed and fully understood by the lessee. For instance, the lessor rarely gives an interest rate for the monthly payments, but such interest can be computed relatively easily using financial mathematics by taking into consideration all the disbursement in the payment schedule. If upfront payments are requested, the total cash flow that results from the lease structure can reduce the initial appeal that seemingly low monthly payments have on the lessee.

The normal set of data that are present in lease contract are: the description of the asset to be leased; the eventual guarantees on the asset; the duration of the lease financing; the price; the deposits, the prepayments, the entity of the monthly payments together with the indication of the interest rates and spreads; the eventual final bullet payment to complete the change of ownership; commissions and fees due to the lessor for setting up and operating the lease financing; and interest rates on nonpaid monthly payments. Some special clauses can be entered on a case-by-case basis, for instance, having in place full-fledged O&M contracts (see Chap. 7), since both leases most often entail the presence of a third party that performs the maintenance and/or the operations of the asset that is leased. In case of assets that have significant capital costs, the lessor can request that the performance of the assets be checked by the lessor itself and, in case of asset low performance, a suitable technical advisor to analyze the causes of low performance with the lessee to bear the duty to repair the causes of low performance according to technical advisor's specifications.

8.7   Project Financing

PF is a financing structure that employs equity and debt for the construction and operations of large infrastructural projects where revenues are secured by strong and binding off-taking contracts or governmental-backed agreements. The parties that are usually involved in a PF are the project company or the *special purpose vehicle* (SPV) company that is created for owning and managing the asset, the advisors (technical, legal, insurance, and tax), the EPC and O&M contractors, the lender parties, the arrangers, the managing bank, and the syndication of other banks providing the loan to the SPV, the account bank where the SPV accounts are held, the off-takers, and the regulatory agencies. A security trustee can also be present to manage the interests of the different groups of lenders or other creditors interested in the PF structuring.

A peculiarity of PF is that the debt is *nonrecourse*, or limited recourse, meaning that in case of default on the part of the borrower, it is not possible for lenders to seek additional compensation apart from the collateral that has secured the loan. The collateral is usually the ownership of the asset that is financed under the PF scheme, and such guarantee is accepted by the lenders even if the collateral value does not fully cover the value of the loaned amount.

Other characteristics of PF that make it appealing to entrepreneurs and investors alike is that the assets are encapsulated in the SPV and are, therefore, maintained off the balance sheets of the companies, together with the possibility to use debt leverage linked to the secure cash-flows of the project, often favorable taxation rules and financing terms, the sharing of risk between the parties, and the differentiation of a portfolio of assets by country and industry.

As a counterpart, structuring PF is not easy and involves a lot of competences which normally imply high development and transaction-specific costs to the developing company. Another disadvantage of PF is that the underlying technology, to be bankable, has to be mature or demonstrated with pilot or demo projects; new technology solutions, that do not yet have a proved business space, are not financed by banks under PF schemes.

The economic viability ad attractiveness of a project under a PF scheme is judged by its IRR (see Chap. 6). Depending on whether or not debt is considered in the analysis, two IRRs are considered in PF:

- the *project IRR*, which is the unlevered IRR of the free cash flows of the project (no debt is used to finance the project)
- the *equity IRR*, which is the levered IRR of the free cash flows of the project

A PF-backed project normally goes through a number of very distinct phases: the development phase, the contracting phase, the financial close, the construction of the facility, its testing and commissioning, its operations, and its final decommissioning. Each of the phases are impacted by a number of specific risks that must be taken into consideration even before committing to the start of development of the project.[4] PF entails the co-participation of equity and debt. While equity is based on the decision of the entrepreneurs, debt is left to the interaction between the entrepreneurs themselves and the bank officers that are assigned to decide on providing the loan or not. Reaching *financial close* creates important risks to the equity owners since closing is not necessarily true at the beginning of development. During the months or years of project development, many conditions can change and modify the original business model on which the construction and operation of the asset was believed to be a sound business idea. Change in law, changes in macroeconomic conditions, country risk, off-taking risks, can come in the way of reaching the COD of the project, delaying it, or making it downright unfeasible.

The development and construction risk is borne by a number of the participants in the PF, each of them sustaining different impacts depending on their respective roles. For example, the firm strives to minimize development costs, to ensure favorable conditions in construction and operations contracts, to secure financing at the earliest possible; lenders on the other hand try to enter the project when risks are minimized and the asset is nearing end of successful construction; finally, contractors need to minimize their costs while at the same time providing the firm with the negotiated build quality and timing. Part of the risks are mitigated by contracts, insurances, and guarantees as discussed in Chap. 6.

Legal and political risk is borne by most of the participants, and the usual outcome is a cause for termination in many PF contracts. Examples of this class of risk can be changes in law, flaws in legal documents, wrong or partial permitting resulting in invalidity of the land and easement rights, corruption, and fraud. Even public opposition can come in the way of the realization of a project. Sovereign guarantees, government comfort letters, support agreements, or loan agreements, can be potentially provided by the local government to ensure that the project company will be compensated if certain events do or do not occur during the life of the project. As a counterpart, a bid bond could be

---

[4] More information on risk analysis and mitigation can be found in Sec. 6.11.

requested by a local government that desires to ensure that the project owner that wins a bid for an infrastructure facility will actually proceed with its construction.

Financial risk is sustained by both the firm that borrows the money and the lenders. While the firm runs the risk of not being able to service the debt (failing to pay interest and repay capital), the lenders run the risk of not receiving debt amortization rates in due time or at all. Lenders hedge their risks by requesting security packages and collaterals on the loan. The banks will ensure that they have security over the sales proceeds, by making sure that money is deposited on reserve accounts and that appropriate clauses in contracts grant the lenders appropriate termination in case some specified parameters are not met during the course of the loan period. Other risks are caused by variations in foreign exchange rates and interest rates; for both, lenders can counteract by hedging with swap contracts.[5] Within the set of guarantees, a *claw-back guarantee* can be enforced in order to ensure that the borrower returns cash distributions to the SPV to the extent required by the project to honor its debt service.

Since a loan is issued by the lenders, the PF contract is similar to a loan contract, and all set of similar covenants, security package, guarantees, control accounts, and project ratios are enforced in a contract for PF (see Sec. 8.5.1).

A typical information memorandum for a PF contract will include, between the others:

- the description of the borrowing entity and sponsors, with information on organization, incorporation, country, and ownership structure
- the description of the project, its location, main technical and economic data, and status of permitting
- the senior debt loan amount
- use of loan proceeds for the advancement of the project
- the identification of the collateral
- the equity terms as contributed by the project sponsors
- the construction budget, the working capital needs, and the handling of eventual cost overruns

---

[5] The use of derivatives helps lenders to protect themselves from a large share of risks. Derivatives can be:

- *Futures* are contracts between two parties to buy or sell an asset for a price agreed today (the *strike price*) with delivery and payment occurring at a specified future date. The *spot price* is the price at which the asset is sold on the spot date. The difference between the spot and the strike price represents the *forward premium* (a profit for the buyer) or the *forward discount* (a loss for the buyer). The purchaser of the underlying asset of the future is said, in derivative jargon, to be "long" and the party that sells the asset is said to be "short."
- *Forward contracts* are contracts similar to futures but traded through *over-the-counter* contracts.
- *Options* (see Sec. 6.9.4).
- *Swaps* are contracts with which the parties exchange cash flows of each other party's financial instruments. For instance, the *interest rate swap* is an exchange of payments of a party having to bear a fixed interest rate loan, while willing to bear a fixed interest loan; this party exchanges cash flows with another party who is paying a fixed interest rate but wants to pay floating interest rate. Another example is a *currency swap* that protects one party from the uncertainty linked to foreign currency when such party is paid, or pays, in a currency not of the party's own country.

- the sponsor guarantees
- the interest rate, the interest rate swap, the debt amortization schedule, and its mechanics
- the fees payable to arrangers, to the lenders
- the election of the technical, legal, insurance, and tax advisors
- the governing law and jurisdiction

8.8   Public Funding

Many can be the financing opportunities made available from the government aimed at helping firms to innovate and upgrade their infrastructures. Apart from the FIT scheme discussed in Sec. 8.2.3 other monetary contributions that are common across countries are:

- capital grants
- cofunding
- interest subsidies

Capital grants are computed on the percentage of expenses that are accepted and validated by the financing organization. After the provision of the capital grant, no redemption of the grant or interest payment are asked back. No guarantees on the receiving firm are normally needed to apply for the grants, the goal of which is to help the entrepreneur, or the firm to acquire machinery, equipment, or deploy infrastructures that have an important and lasting impact on the firm's operations and its development in the future. Capital grants are usually paid on exhibition of expenses receipts.

Cofunding is the provision of debt, for all or part of the total consideration needed by the borrower, from public organizations to firms at an interest rate lower than market standard conditions.

Interest subsidies are contributions made on the interests paid by the borrower during debt servicing of a loan. The interest subsidy is paid back to the borrower and is computed on the same loan, or a loan with same amount but different maturity date, with an interest rate decided by the subsidy scheme. In practice, the borrower enters a loan at available market rates and conditions, and subsequently apply for the interest subsidy. Since a loan has already been granted by a financing institution, no further guarantees are requested for the interest subsidy.

Governments, or supranational organizations (like the European Community), can also fund new ventures or small medium enterprises (SMEs) by providing grants under the scheme of research and development frameworks. Normally, SMEs can enter into consortium agreements with partners like other SMEs, universities, larger organizations, and apply for funding on a competitive basis to secure a portion of the total amount of funds that are reserved each year. The application process usually follows the steps of:

- finding the partners
- submitting the proposal

- evaluation by experts
- grant agreement

Application to grants can be time consuming and the outcome very uncertain due to the competition between many different consortia that can be applying for the same grants.

Funds can also be granted as loans by supranational organization like the World Bank to develop infrastructure projects all around the world by managing loan programs and providing guarantees or indirect support. Another form of governmental support to firm are the *export credit agencies* that arrange direct lending or provide international financial guarantees helping firms in importing or exporting their goods in the international marketplace.

References

1. Vance D. E. (2003). Financial Analysis and Decision Making. McGraw-Hill.
2. Eydeland A., Wolyniec K. (2003). Energy and Power Risk Management—New Developments in Modeling, Pricing, and Hedging. Wiley.
3. Gatti S. (2008). Project Finance in Theory and Practice—Designing, Structuring, and Financing Private and Public Projects. Elsevier.

Conclusion

Hopefully you have reached this point of the book, where I have the pleasure to share with you some final thoughts.

This book has tried to convey the message that the implementation of energy storage with clean energy power plants may really make the difference to the energy economy of our world.

This book has also given you many tools to be an active member of this shift toward such a new paradigm of a decentralized, democratic, cheap, and widespread energy economy.

So many are the opportunities, the advantages, the benefits to our world that they should not be neglected, and decisions should not be based only on mere economic and financial analysis.

Green electric energy storage is already a valid and viable technology, and current and future advancements will further increase its efficiency and effectiveness.

Now we need to start committing to this next phase of progress and value the incredible possibilities that lie in front of our eyes.

It will be fun to see when this happens in the near future. I hope we will have fun together.

All the best,
Gabriele

Index

M

N

CPSIA information can be obtained at www.ICGtesting.com
Printed in the USA
BVOW03*0939290816

460185BV00003B/4/P